EXPERIMENTAL FUN
WITH THE YO-YO
—and Other
Science Projects

Other books by Al G. Renner

How to Build a Better Mousetrap Car—and
Other Experimental Science Fun

How to Make and Use an Electric Motor

How to Make and Use a Microlab

Tumblers' Manual
(with Ralph LaPorte)

EXPERIMENTAL FUN WITH THE YO-YO

—and Other Science Projects

Written and illustrated by
AL G. RENNER

DODD, MEAD & COMPANY
New York

6|79

Library of Congress Cataloging in Publication Data

Renner, Al G
    Experimental fun with the yo-yo and other science
projects.

    Includes index.
    SUMMARY: Presents science experiments based on a yo-yo,
wind wagon, cork sailboat, inverted tumbler, Cartesian
diver, and hang glider which the experimenter makes him-
self and then changes as he tests scientific data.
    1. Science—Experiments—Juvenile literature.
2. Yo-yos—Juvenile literature. [1. Science—Experiments. 2. Experiments]   I. Title.   II. Title: Science projects.
Q164.R45   530'.028   78-23569
ISBN 0-396-07657-2

CONTENTS

INTRODUCTION

The Spinner

This is a book for young experimenters. If you are one of those people who likes to find out things for yourself, you will enjoy it. You will learn some of the skills and techniques which make our scientists such successful problem solvers.

Suggestions for making the models will be given. However, instead of putting the finished models on the mantle for display, you will be encouraged to change them again and again.

You are invited to try and to fail as well as to succeed in your experimenting. Most of all, it is hoped that you will have fun in the things that you do.

Experimenting is simply making changes purposely so that you can learn more about things. You will design and test your own theories about what makes your models work. When you understand and when you can put all of the best properties and the best conditions together, you can make that super, very original model that you will be proud to display in your room. Best of all, you will become a capable and an independent experimenter who can solve your own problems. It is a nice feeling to know that you can take care of yourself. You will feel more self-reliant and grown-up. Have experimental fun.

HISTORY OF THE YO-YO

The laughter of Chinese children, gaily spinning their yo-yos, was heard over two thousand years ago. A close look at their gyroscoping toys would have revealed that they were made of heavy metal or bone buttons wired through the holes and twisted together.

The yo-yo followed Marco Polo from China to Greece where it became very popular. Then crusaders took it from the Near East back to the royal courts of Europe. Soon the sport had spread throughout the countryside. Mothers must have complained about missing coat buttons.

From China the yo-yo also spread eastward, to the delight of children on the Philippine Islands. It became more popular there than any place in the world. It was even developed as a weapon for jungle warfare. A rock yo-yo has often downed an enemy when spun from the practiced hand of a soldier hidden on the limb of a jungle tree.

In the 1920s the late Donald Duncan was entranced by the spinning discs when he saw Filipinos with them on the docks in San Francisco. He registered the native Philippine name, and the yo-yo has been called that ever since.

The yo-yo has been used as a jungle weapon

THE YO-YO PROBLEM

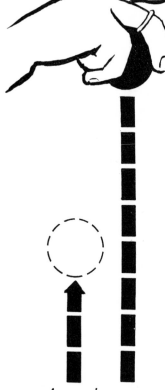

How many kinds of yo-yos can you make? If you know what makes a yo-yo work, you will be surprised at the number of things that will yo-yo for you.

If you can spin a store yo-yo well enough to do the fundamental stunts which are illustrated on the next two pages, you will be able to do more experimenting. Note that there are two kinds of stunts. Start with the string tied tightly to the axle.

A good yo-yo should return about halfway up the string all by itself when dropped from the hand.

No yo-yo will return to your hand all by itself. The first thing that you must do if you wish to be a good yo-yo experimenter is to add just a little extra energy to the spin with a flick of your wrist as you cast the yo-yo. You can add another slight bit of energy by making it "crack the whip" with a slight tug as it rounds the bottom end of the string so that it climbs to your hand.

BASIC YO-YO STUNTS
with a tight loop on the axle

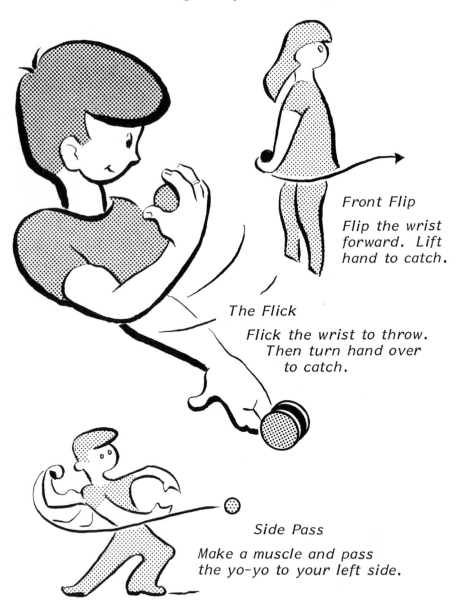

Front Flip

Flip the wrist forward. Lift hand to catch.

The Flick

Flick the wrist to throw. Then turn hand over to catch.

Side Pass

Make a muscle and pass the yo-yo to your left side.

BASIC YO-YO STUNTS
with a loose loop around the axle

Tie a loose loop with a bowline knot or place the yo-yo in the loop of the doubled string. When you use a loose loop around the axle, it will unwind down to the end of the string and sleep or spin in the open loop. Jerking the string will jam the loop so that the yo-yo can roll back to your hand.

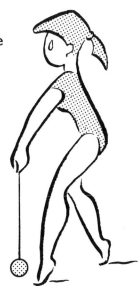

The Spinner

How long can you make it sleep at the end of the string?

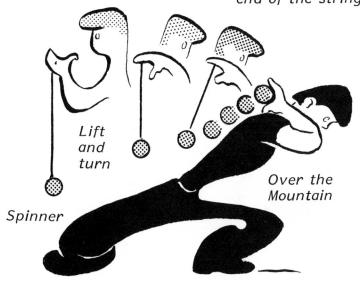

Lift and turn

Spinner

Over the Mountain

11

HAVE A PURPOSE FOR YOUR EXPERIMENT

Experiments are purposeful changes which you will make to find out more about your yo-yos and the other problems in this book. Your changes can have two purposes. You must decide which purpose you have in mind before you start each experiment for it to be most meaningful.

Experiments can generate new data. You will change and keep changing the parts of your yo-yos, for example, to discover new data. Data are the true facts or the bits of information which you discover. You will learn what your yo-yos really do rather than what you think that they are doing. You will experiment most often by changing a condition; but you can change the materials of which the parts are made; and you can even substitute new parts.

You can use experiments to test your new inferences, hypotheses, and theories. These words refer to levels of understanding. No matter what you call your own ideas, the most important thing is that you design new experiments immediately to test them.

Yo-yo as you walk

TEST YOUR THEORIES WITH PREDICTIONS

Your theories are your invented ideas. Theories are useful devices of the mind which can help to control, to explain, and to predict how things will work. Just change a condition and predict how the yo-yo will act differently before you throw it.

Your prediction might be, "If I change to a yo-yo with only one wheel, I predict that it will tip to the side with the wheel." Your yo-yo will tell you how good your understanding of its action is.

Don't worry if your predictions are wrong at first. You really will be closer to the truth because you will know at least what does not work. You will be much closer to the truth than if you did not try to predict.

If you predict and find yourself right several times, feel good because you are beginning to understand the problem even though you may feel that you cannot explain it yet. If you predict and find yourself wrong, feel good also because you know now what not to do. Remember, make a prediction each time that you set up a new experiment.

Yo-yo as you run

CHANGE THE STRING

Change the length of the string. Try different lengths out of a second-story window. What is the limit for the length of a yo-yo string? Which length works best? Your own yo-yo will tell you what you want to know when you learn to ask it very specific questions.

Change the material of the string. Does it have to be string? Could it be thread? How about cord? Mylar recording tape? Be sure that you try nylon. Make an elastic string by looping narrow rubber bands together. Your mother just might even have a long elastic cord in her sewing kit.

Change the surface of your string by waxing and even lubricating it. Remember that you can twist your string to any strength or to any width when it is doubled. Be sure to change the kinds of loops for your finger and for the yo-yo axle. What new experiments can you design?

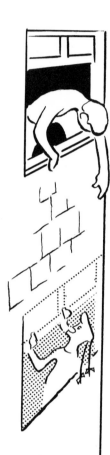

"Hey, watch what you are doing!"

14

KEEP CONTROL

To keep control of an experiment means that you should change, or manipulate, only one variable at a time. If you change more than one condition and something spectacular happens, you never know which condition caused the new action. Start out right by changing the conditions one at a time. You will do that eventually anyway.

For an example, let us imagine that you changed your yo-yo shape to a square piece of a new material. Then you painted this yo-yo with a purple paint that a friend gave you. You drilled a hole through each corner because it seemed like a good idea. You promptly changed your mind and packed the holes with bubble gum. Then you melted a candle in a metal cup and soaked the yo-yo in the warm wax to make it smoother.

You got a surprise when you started to spin your yo-yo the next day. It whistled a quiet tune as you warmed it up in practice. Then suddenly, it glowed bright red and gave off a puff of purple smoke just before it exploded in a brilliant green flash from the end of the string.

What caused this interesting display? You will never know until you start all over again to change just one condition at a time so that you can tell which made the difference in the yo-yo's performance. You might as well start with all conditions and properties under control in each experiment and save yourself much time.

Walking the dog

CHANGE THE STRING

A bicycle wheel gives you the feel of a yo-yo

Did you ever hold a spinning bicycle wheel on only one side? It is an experience which you should have to understand the yo-yo. Hold the wheel by the two short axle ends with your arms outstretched in front of you. Have a friend spin the wheel as fast as possible. Then remove one supporting hand. You will be amazed that the wheel does not lose its balance immediately. Expect it to lose its position as the spinning motion is slowed by friction.

Hold the spinning wheel and try to walk around a table, turning very sharp corners. Walk straight, and then try to reverse and go in the opposite direction, holding the wheel upright. What you feel is the action of a gyroscope. Once you put it in motion, it has a tendency to stay in the same vertical position as long as it is spinning.

16

CHANGE THE STRING

Change the position of the string on the axle. Who said that it should be in the middle? Make a wide-channel yo-yo with a pencil and some wooden wheels.

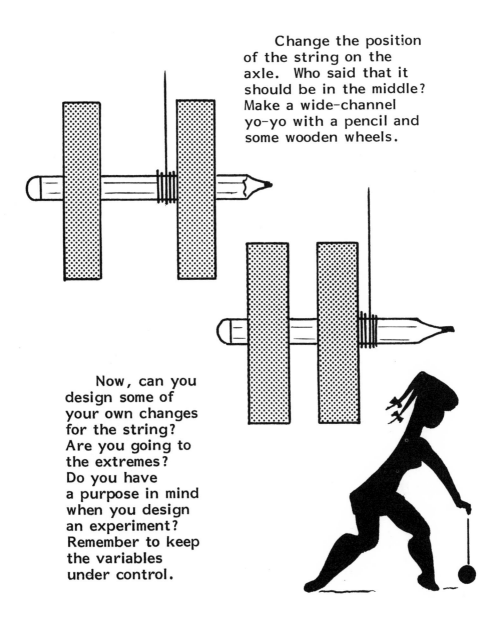

Now, can you design some of your own changes for the string? Are you going to the extremes? Do you have a purpose in mind when you design an experiment? Remember to keep the variables under control.

CHANGE THE STRING

Would two strings work better than one string? Would they give you better balance? If they did, you could use a ball, a disc, a wheel, or a spindle for your yo-yo. Why don't you try and see?

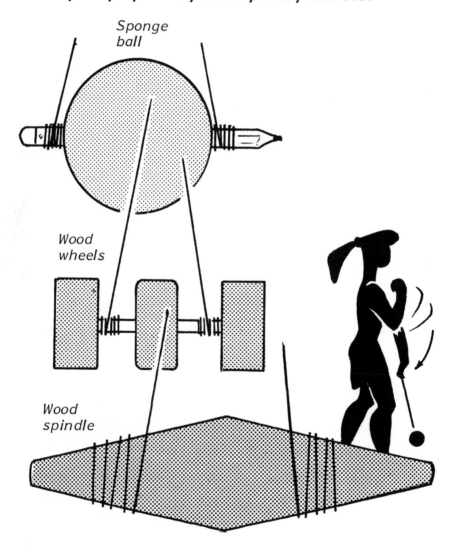

Sponge ball

Wood wheels

Wood spindle

CHANGE THE STRING

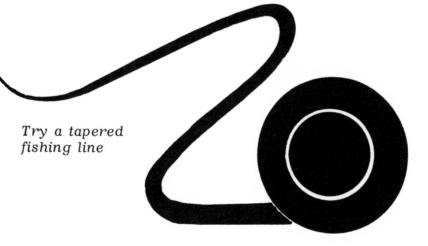

Would a tapered string help the yo-yo spin faster? Could the wound-up coil of string be thought of as part of the axle? If so, it is a larger axle at first, and the yo-yo spins fast as it starts to roll downward. As the string unrolls, the axle becomes smaller, and the yo-yo does not spin as fast as it could.

Would a tapered string spin the yo-yo faster? Toward which end should the string be tapered? You can make a tapered string by coiling nylon thread around the original yo-yo string. Tie an overhand knot every tenth turn.

Change the shape of the string from round to flat by using cloth tape or Mylar recording tape. These wind around the axle very easily.

Try a tapered fishing line

CHANGE THE SHAPE

Does a yo-yo have
to be round? Could this
be just for the convenience
of fitting it to the hand?

Saw two squares off
a three-inch-wide board.
Don't be concerned if
they are not exactly
square.

Sand the corners and
edges round so that the
string does not catch.

Lay two one-eighth-
inch spacers on one of
the yo-yo halves.

Lay the second half
of the yo-yo in place.
Draw a line from corner
to corner. Pound four
box nails at equal
distances from the
center for the yo-yo
axle.

Put a string on
your yo-yo and have
experimental fun.

What other changes
in shape can you design?

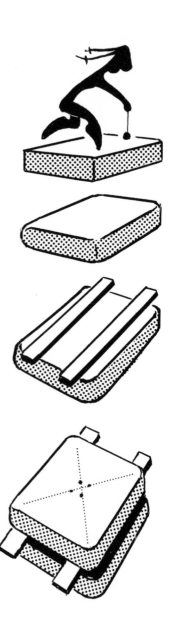

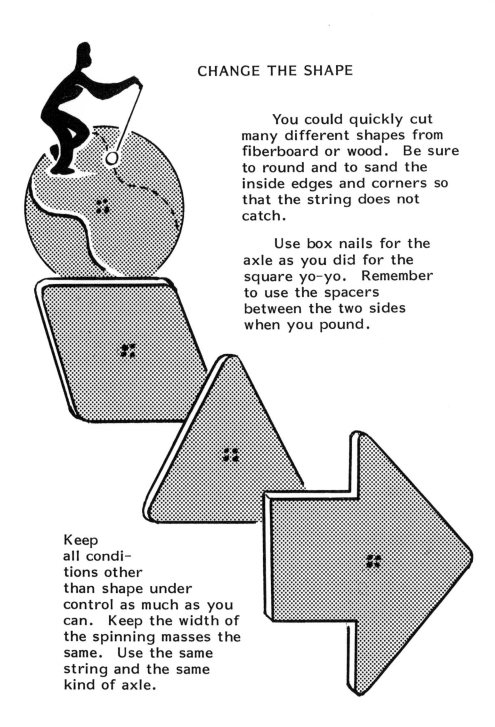

CHANGE THE SHAPE

You could quickly cut many different shapes from fiberboard or wood. Be sure to round and to sand the inside edges and corners so that the string does not catch.

Use box nails for the axle as you did for the square yo-yo. Remember to use the spacers between the two sides when you pound.

Keep all conditions other than shape under control as much as you can. Keep the width of the spinning masses the same. Use the same string and the same kind of axle.

CHANGE THE SHAPE

You can find many objects with different shapes that will spin as yo-yos if you look around your house.

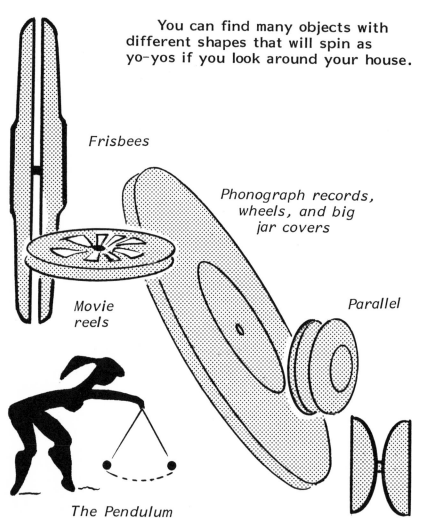

Frisbees

Phonograph records, wheels, and big jar covers

Movie reels

Parallel

The Pendulum

Butterfly

It just happens that standard yo-yos count off seconds with their swings. Flick a spinner, start it swinging, and count.

CHANGE THE SHAPE

Does it make a difference when you change the shape? Ideally, to keep control of your experiment, every shape should be made of the same material. Every shape should be the same size and the same weight. This can get to be very expensive; and young experimenters cannot afford to keep this much control.

Therefore, when you find a shape that does make a difference in the spin, you must always ask yourself, "Could it possibly be the weight of the shape, the size of the shape, or the friction caused by the shape that made the difference?"

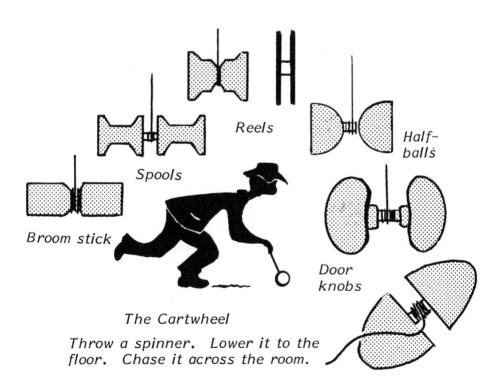

Reels

Half-balls

Spools

Broom stick

Door knobs

The Cartwheel

Throw a spinner. Lower it to the floor. Chase it across the room.

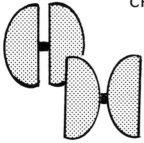

CHANGE THE SHAPE

"Which shape is best," the argument always goes, "a parallel or a butterfly for the yo-yo?"

"Best for what?" you should ask. Could it be that both have some advantages and disadvantages?

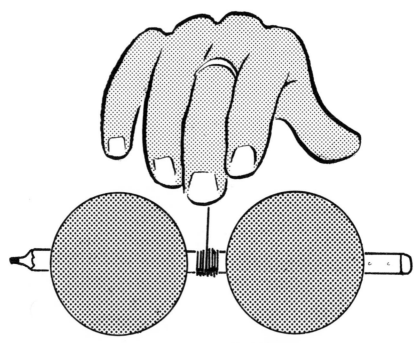

Sponge-ball yo-yo

Stick a pencil through two sponge balls or through two old tennis balls. Would two apples or two oranges yo-yo as well as balls? How about potatoes?

CHANGE THE SHAPE

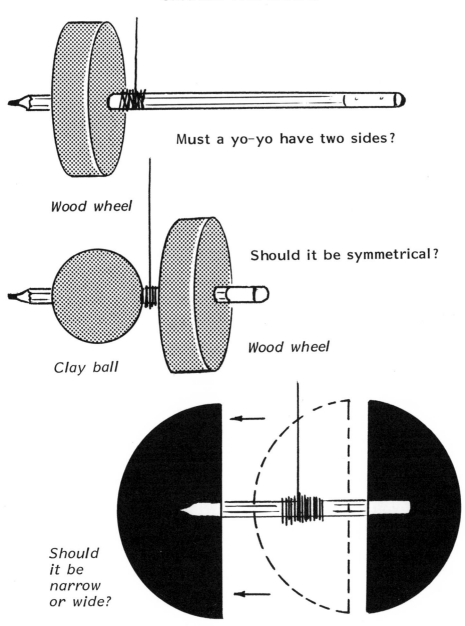

Must a yo-yo have two sides?

Wood wheel

Should it be symmetrical?

Clay ball

Wood wheel

Should it be narrow or wide?

GO TO EXTREMES

Go to extremes when you want to know if a new condition or a new property will make a difference. This will soon tell you the limits of your materials.

A condition is a state of being, one of the more-or-less things that you can change. A property is a distinguishing characteristic. It cannot be changed, but you can insert a new property into the experiment by trying a new material.

What is the smallest yo-yo that you can spin?

For example, in yo-yo experimenting you should go to extremes when changing the length of the string. Don't just make the string a little longer or a little shorter than the one which was on your yo-yo. Try an extremely long string to spin your yo-yo out of a second-story window. Then cut the string as short as you can to try again.

If you start with a featherweight foam plastic yo-yo, it is a good idea to go next to a high density material such as iron or lead. After you try a small button yo-yo, you should go to the other extreme of size to try the largest disc or wheel that you can spin.

GO TO EXTREMES

Some of the conditions which you should take to extremes are listed below. It really does not matter with which extreme you start. Just be sure that you can afford the change and that it is not dangerous.

Smallest	–	Largest
Smoothest	–	Roughest
Lightest	–	Heaviest
Narrowest	–	Widest
Slowest	–	Fastest
Shortest	–	Longest
Weakest	–	Strongest
Thinnest	–	Thickest
Softest	–	Hardest

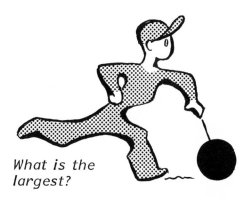

What is the largest?

CHANGE THE AXLE

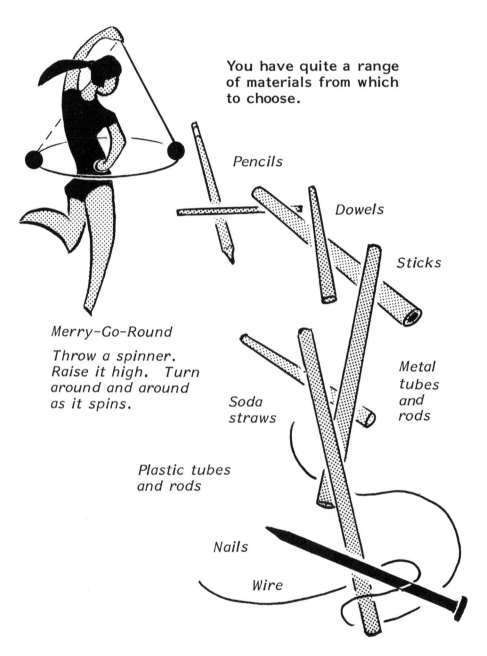

You have quite a range
of materials from which
to choose.

Pencils

Dowels

Sticks

Merry-Go-Round

*Throw a spinner.
Raise it high.   Turn
around and around
as it spins.*

Soda
straws

Metal
tubes
and
rods

Plastic tubes
and rods

Nails

Wire

CHANGE THE AXLE

The first yo-yos in Europe were heavy buttons held together with copper wire axles. Wire axles will enable you to make a yo-yo more quickly.

The Buzzer

Flick a spinner. Lower it to the edge of a paper or cardboard which will make a buzzing sound. How long can you keep it spinning?

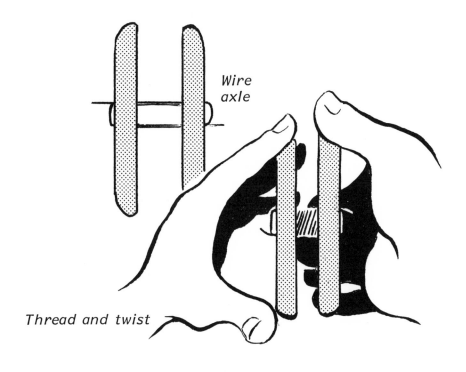

Wire axle

Thread and twist

CHANGE THE AXLE

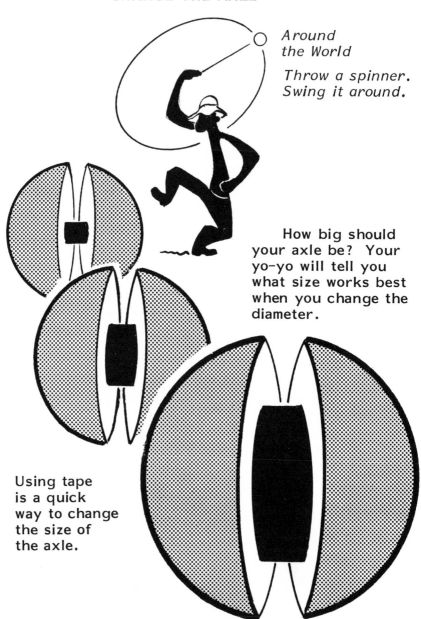

*Around
the World*

*Throw a spinner.
Swing it around.*

How big should
your axle be? Your
yo-yo will tell you
what size works best
when you change the
diameter.

Using tape
is a quick
way to change
the size of
the axle.

CHANGE THE WEIGHT

Make a foam yo-yo first. Compare its action to that of a wooden model of the same size and shape. You probably will have to improvise the heavier models.

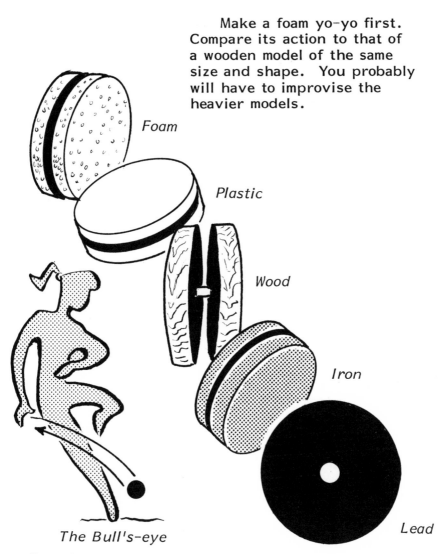

Foam

Plastic

Wood

Iron

Lead

The Bull's-eye

From how many positions can you hit the bull's-eye?

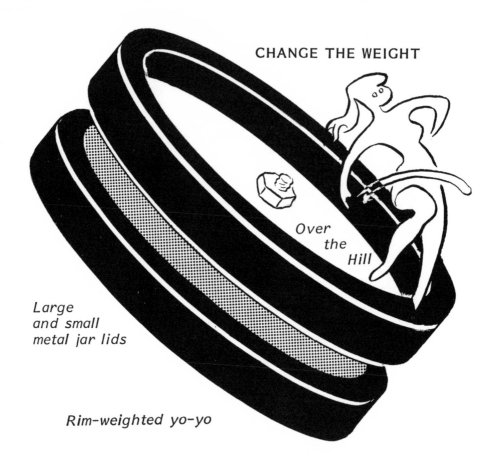

CHANGE THE WEIGHT

Over the Hill

Large and small metal jar lids

Rim-weighted yo-yo

Make this rim-weighted yo-yo and the center-weighted yo-yo on the next page with conditions as much under control as you can. Make this rim-weighted yo-yo first. Place a small jar top inside a large jar top for each yo-yo side. Fasten the rim-weighted yo-yo with a small bolt. Use an extra nut as a spacer between the two sides. Fill the space between the two rims with wax, plaster, or modeling clay.

CHANGE THE WEIGHT

To make the center-weighted yo-yo, drill a larger hole for the new axle. Use a very large bolt to secure the two sides together this time. Remember to get an extra nut which you can use for the spacer between the two halves. Remove the rim-weighted packing. Try to make the large bolt and nuts equal the weight of the packing plus the small bolt and nuts as closely as you can.

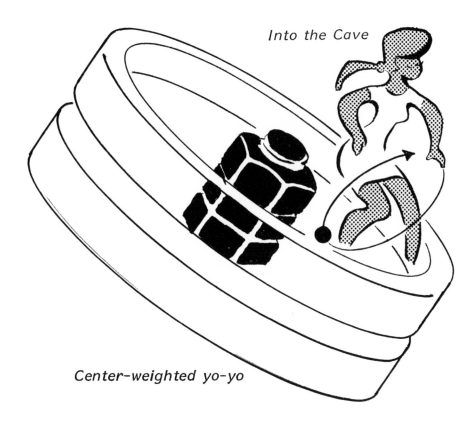

Into the Cave

Center-weighted yo-yo

CHANGE THE WEIGHT

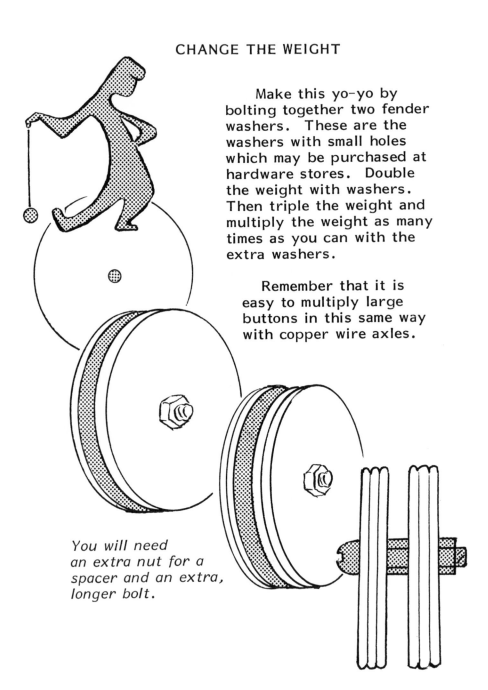

Make this yo-yo by bolting together two fender washers. These are the washers with small holes which may be purchased at hardware stores. Double the weight with washers. Then triple the weight and multiply the weight as many times as you can with the extra washers.

Remember that it is easy to multiply large buttons in this same way with copper wire axles.

You will need an extra nut for a spacer and an extra, longer bolt.

RECORD EXPERIMENTAL DATA

Scientists constantly record data and their own thoughts. Their thinking is revised often, but their data is never changed. Make your own experiment records brief, yet accurate. Laborious record keeping can take away the joy of experimenting. Records should help rather than hamper your thinking.

Scientists are in the habit of using abbreviations, symbols, formulas, equations, sketches, and shadow drawings to record ideas. Once a word is used, it can be abbreviated when used again. Science data is seldom recorded in narrative or story form. Paragraphs and even sentences are a handicap. Record your data with short phrases of well-chosen words. Laboratory records are written differently than English compositions are.

Measure your data as accurately as possible with the metric system. Measure distances in millimeters, centimeters, and meters; measure weight in grams and kilograms; and measure volumes in milliliters and liters. Measurements make the sensing of data much more accurate. All scientists have used the metric system ever since it was originated.

Record the data immediately after the experiment

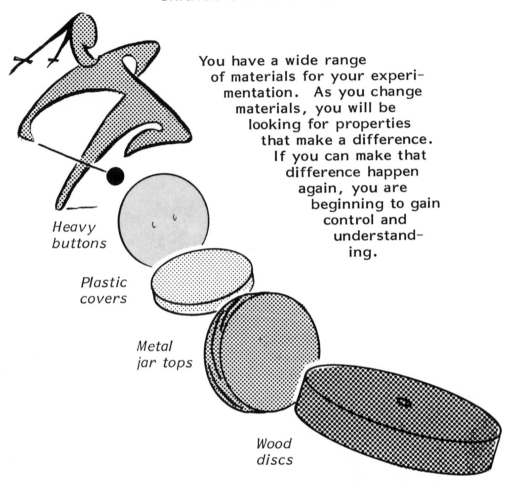

You have a wide range
of materials for your experi-
mentation. As you change
materials, you will be
looking for properties
that make a difference.
If you can make that
difference happen
again, you are
beginning to gain
control and
understand-
ing.

Heavy
buttons

Plastic
covers

Metal
jar tops

Wood
discs

You will want to change some materials by sand-
ing, cutting, or sawing. Other materials you will
change by casting into new shapes. One thing you
can be sure of, you won't be doing much experimenting unless you can change your materials and
the conditions under which they exist.

CHANGE THE MATERIAL

Remember that you can always substitute new objects with new properties. Properties are the identifying characteristics of materials. They are the attributes which never change. Lead metal is always heavy and dense. Aluminum metal will always be light and less dense. You will not be able to change the properties of materials in your experimenting. You can always substitute new materials with new properties.

It would be well to check periodically to see if you are still working like a scientist. Do you have a purpose for each experiment? Are you still going to extremes when you change conditions? Are you recording your data promptly? Are you making predictions about what will happen to test your ideas before each experiment starts? Do you still ask specific questions when you want your models to give you data?

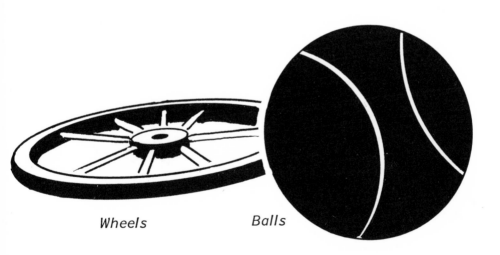

Wheels   *Balls*

CHANGE THE SIZE

Can you spin a small button yo-yo on a thread? Can you make a large bicycle-wheel yo-yo work?

A group of engineering students at the Massachusetts Institute of Technology made two bicycle wheels yo-yo up and down a twenty-one-story skyscraper on a 265-foot cord.

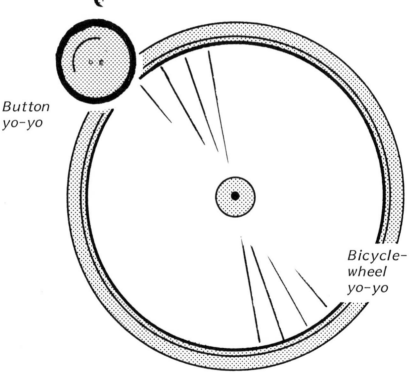

Button yo-yo

Bicycle-wheel yo-yo

HOW TO TEST YO-YOS

Tape a meter stick to a wall to test the roll-back quality of your yo-yos. Tape a coat hanger wire hook at the top of the meter stick. Put a small clamp on the hook.

When a yo-yo is to be tested, let it hang on a meter of string so that it almost touches the floor. Then wind it up to the clamp.

Now, let it drop to unwind by itself and return up the string. Mark with a tape how high it climbed.

A good yo-yo will return at least 50 percent of its string length. A return of 75 percent is outstanding.

When you wish to buy a yo-yo at a store, tie the string to the axle tightly and hang the yo-yo from belt level to see how much return you get.

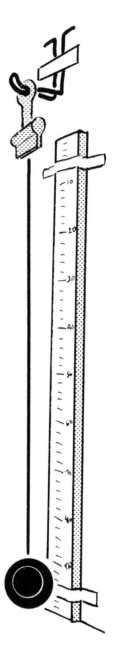

Meter stick taped to wall

DON'T BE AFRAID TO TRY AND FAIL

Scientists learn much from experimental failures. When NASA was first planning the trips to the moon, weight limitations were important. They did not have rockets which were powerful enough to boost all the equipment which they wanted to place on the moon. Many materials were tested to fail purposely. Each part had to be as light as possible and yet strong enough to do its job.

You, too, can learn much from failures if you try. Think of that little old lady who lives down the street. She takes great pride in the fact that she has not had an accident or even a traffic ticket for many years. She drives two blocks down the street to her market each week avoiding the peak traffic. Is she really a good driver?

Try again

This little old lady would probably have been a much better driver today if she had been able to drive in extremes of conditions and to have had a few accidents. Imagine her driving and getting stuck in desert sands, bouncing over rocky mountain roads, surging through stormy flood waters, sliding off an icy road, or just speeding down a bustling freeway. She might have collected a few dents in her fenders and even have made a few appearances in traffic court, but she definitely would have had a chance to become a much better driver.

CHANGE TO A DIABLO

The diablo is a cousin of the yo-yo

The diablo is a form of the yo-yo which has come to us from South America and China. It is more like a spool and does not have the string attached.

Originally it may have been cut from a tree branch and shaped like a spool. You can learn much about your yo-yo by experimenting with the diablo.

Cut your first diablo from a large tree branch. Are narrow ones better than wide ones? Are short ones better than long ones? Your diablos will tell you. Improvise diablos from large spools and door knobs. Also stick a pencil through two foam cups and fill them with clay, wax, or plaster.

CHANGE TO THE DIABLO

Start your diablo spinning by rolling it on the floor. Lift the rolling diablo with the string on sticks. Keep it spinning with short, jerky, lifting movements. The faster you make it spin, the better its balance will be. Keeping the string in line with the spool will also help to keep it in balance.

Tree branch

Door knobs

What new diablo experiments can you design?

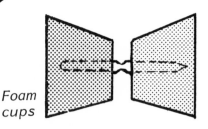

Foam cups

BASIC DIABLO STUNTS

How fast can you make it spin? Can you make it spin backward as well as forward? Can you start it spinning without rolling it on the floor?

The Spinner

The Toss-up

Can you catch it again and keep it spinning?

After the step-through, can you step back and start combinations of these stunts?

The Step-through

YOUR SUPER YO-YO

And now, we would like to suggest that you design your super yo-yo. Use the best materials with outstanding yo-yo properties and use them under the very best conditions.

Perhaps you are asking, "Best for what?" This is good. Have a special purpose in mind. Do you want a good spinner? Are you in need of a yo-yo which returns well? Do you want some kind of a novelty yo-yo that hums or plays a tune? Perhaps you need an all-around yo-yo for competition.

Once you decide on the kind of yo-yo that you want, put all of your best experiences and ideas together to accomplish that purpose.

Your super yo-yo, what will it be like?

MATERIALS NEEDED

Bodies
- Paper towel rolls
- Toilet paper rolls
- Yogurt tubes
- Mailing tubes
- Urethane rods
- Soda straws
- Pencils

Sails
- Paper
- Aluminum foil
- Saran Wrap
- Sheet plastic
- Paper

Masts
- Pencils
- Soda straws
- Swab sticks

Wind
- Fan
- Improvised wind tunnel

Wheels
- Yogurt pushups
- Checkers
- Bottle tops
- Plastic jar covers
- Liners from covers
- Cross sections of urethane rods and broom sticks

Axles and bearings
- Coat hanger wire
- Plastic rods
- Small soda straw axles
- Large soda straw bearings

Miscellaneous
- Pins
- Glue
- Toothpicks
- Rubber bands
- Hacksaw blades
- Knife or razor blade
- Thread and string

Blustery weekend days often reveal wind wagons sailing across large, unused parking lots. These land rovers are also known as sail carts and land yachts. Windy days bring them to California dry lake beds and sand dune areas. Surfers on skateboards compete with the wind wagons by spreading their jackets or by holding improvised sails. Building a model wind wagon can teach you much about riding a large one.

CHANGE THE BODY

What shape is best for the body? Could part of it be cut away as unnecessary?

What material is best? What size? Shape? Should the body be long or short? Wide or narrow?

Would heavy ballast in the body keep the wind wagon from blowing over? Would it slow the wind wagon? When you add the ballast, your wind wagon will tell you what is best.

Which is best for securing the mast: glue, tape, or pins?

Soda straws

Paper tube

Sponge

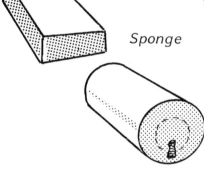

Aluminum can

Desk ruler

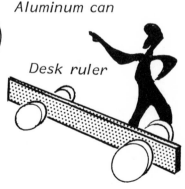

What new bodies can you improvise?

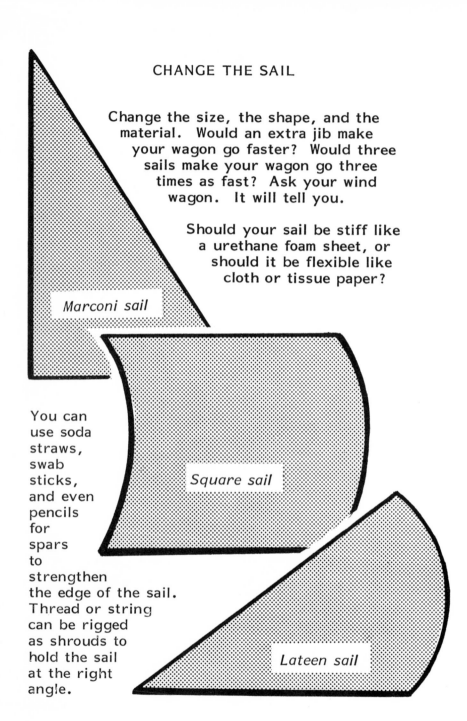

CHANGE THE SAIL

Change the size, the shape, and the material. Would an extra jib make your wagon go faster? Would three sails make your wagon go three times as fast? Ask your wind wagon. It will tell you.

Should your sail be stiff like a urethane foam sheet, or should it be flexible like cloth or tissue paper?

Marconi sail

You can use soda straws, swab sticks, and even pencils for spars to strengthen the edge of the sail. Thread or string can be rigged as shrouds to hold the sail at the right angle.

Square sail

Lateen sail

CHANGE THE WHEELS

A thick urethane or styrofoam plastic rod can be a quick supply of wheels to get you started. You can also cut wheels from a foam plastic sheet if you can draw a good circle with a compass.

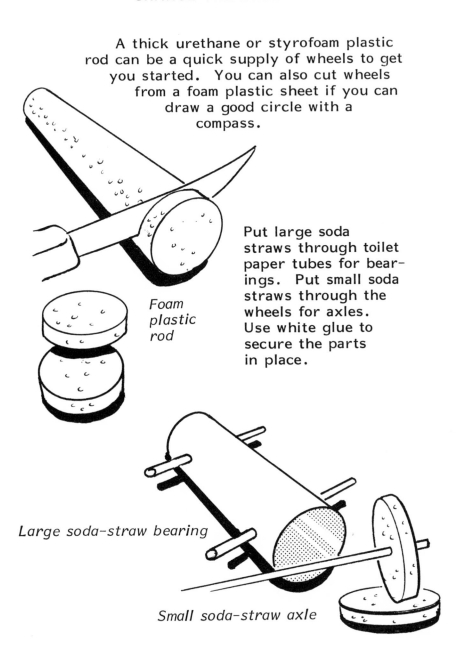

Foam plastic rod

Put large soda straws through toilet paper tubes for bearings. Put small soda straws through the wheels for axles. Use white glue to secure the parts in place.

Large soda-straw bearing

Small soda-straw axle

CHANGE THE AXLE

One of the best wheel-and-axle combinations for small vehicles is the plastic pushup from the yogurt and sherbet which is sold in tubes. This becomes an excellent wheel and axle when used in a soda straw bearing.

Should the axles be short or long? Does it make a difference in the guidance? Always use straw axles of small diameter so that they will fit into soda straw bearings of a larger size.

Straight coat hanger wire makes good axles if you need long ones. Nails can be used to hold each wheel to wooden bodies and sponge bodies. When using narrow wheels such as cardboard bottle top liners or plastic snap-on tops, a wide hub may be needed to keep the wheel in line.

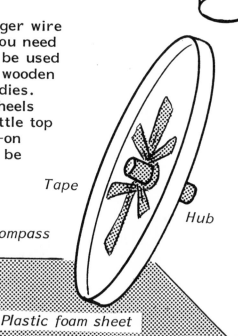

Yogurt

Pushup wheels

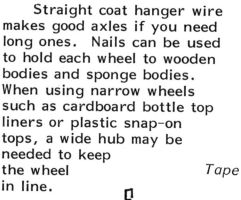

Tape

Hub

Compass

Plastic foam sheet

WINDY DAY FUN

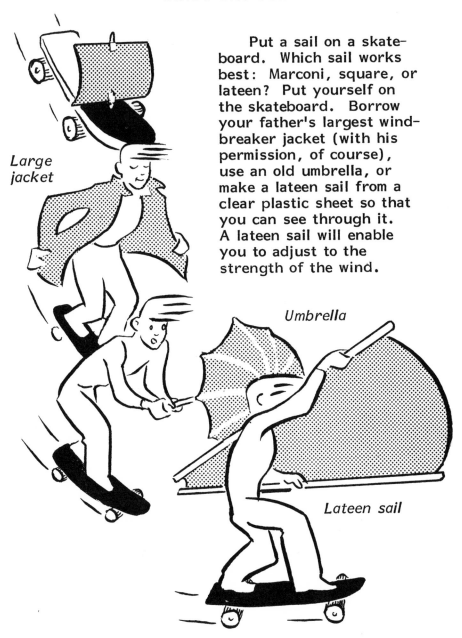

Put a sail on a skateboard. Which sail works best: Marconi, square, or lateen? Put yourself on the skateboard. Borrow your father's largest windbreaker jacket (with his permission, of course), use an old umbrella, or make a lateen sail from a clear plastic sheet so that you can see through it. A lateen sail will enable you to adjust to the strength of the wind.

Large jacket

Umbrella

Lateen sail

If you can build a small model wind wagon, you can build a large one. Ice skates work better than wheels in winter time.

Bicycle-wheel wind wagons sail well on unused supermarket parking lots on windy days.

Ice skates

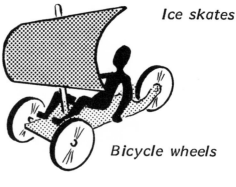

Bicycle wheels

Balloon tires roll wind wagons over California dry lakes and sand dunes.

Balloon tires

EXPERIMENTAL
FUN WITH THE

CORK
SAILBOAT

Main sail

Jib

Mast

Stern   Beam   Bow

Keel

CAN YOU MAKE
A SAILBOAT THAT
WILL TACK INTO THE WIND?

MATERIALS NEEDED

Hulls
  Thermos corks
  Balsa wood
  Foam plastic
  Small rubber bands

Masts
  Toothpicks
  Swab sticks
  Paper-clip wire

Sails
  Typing paper
  Wax paper
  Saran Wrap
  Aluminum foil

Keels
  Lead foil
  Bottle seals
  Paper-clip wire
  Copper wire
  Razor blades with
    single edge

Rudder
  Lead foil
  Aluminum foil
  Paper-clip wire

Wind
  Blow with mouth
  Large fan

It is highly recommended that you start with cork hulls because they are so easy to cut and to shape. However, in your following experiments you should investigate the properties of many kinds of materials. Making small boats will enable you to do much more experimenting. Later on you can make large models and even put radio controls in them.

Many experimenters have made pools for their boats with plastic sheets. The best plastic sheets are the inexpensive tube tents which are used by back-packers. These plastic pools can easily be set up in schoolrooms to experiment with boats.

START WITH A SQUARE-RIGGER

Start with a square-rigger

If you would like to understand sailboats better, you should build some models and experiment with them. Some day you may want to own a sailboat. This experimenting will help you to take better care of your boat, and it will help you to sail it faster and more skillfully.

Make an old-fashioned square-rigger first. You will be able to make it sail with the wind immediately. If it doesn't go straight at first, start experimenting. Start changing things until it goes where you want it to go. Then you can make other kinds of boats to compare with it.

You can start your experimenting in a large bowl or in a pan of water. You can use a wash basin or the bathtub. You will do more experimenting if your boats do not travel far. Just blow with your mouth at first to make a wind. Later you can work on a larger pool with more wind. You can make a good, steady wind with a fan for indoor experimenting.

Wise experimenters start their changes with a split hull. When the two halves of the hull are held together with a rubber band or two, it makes it easy to change the position and kind of mast, keel, and rudder. When you decide which combination of parts works best, you can glue them into place.

CHANGE THE HULL

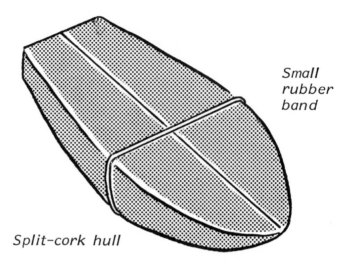

Small rubber band

Split-cork hull

Change the material of which the hull is made. Start experimenting with half thermos corks for hulls, but be sure to change to plastic foam, balsa wood, pine wood, and as many other materials as you can obtain. Don't forget to split the cork hull down the middle so that you can adjust the mast, keel, and rudder more easily. This is important.

Change the size of the hull. Glue two thermos corks end to end and split them the long way for hulls. Plastic foam will enable you to make the quickest changes in size. Make midget models first. Then make a large model with the largest piece of foam that you can obtain. You can always make intermediate-size models by cutting down the large hull.

Change the shape of the hull. Both cork and plastic are excellent materials for changing the shape of your sailboat.

CHANGE THE HULL

Must the hull be sleek and beautifully curved to be fast? Would a raft-like or sled-like flattie sail as well? Your boats will tell you when you change the shape of the hull.

If your mast tips over in the wind, you can try the solutions of natives who live on windy and stormy seas. An outrigger is an outside float which steadies the sailboat. A catamaran is a double-hulled or split-hulled craft. These boats do not require a heavy keel.

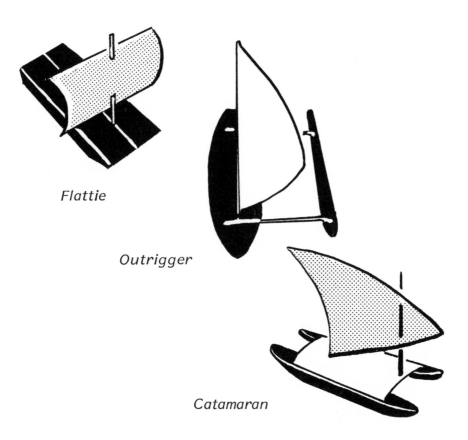

Flattie

Outrigger

Catamaran

CHANGE THE MAST

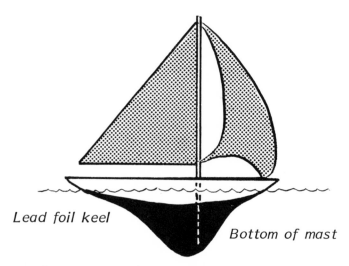

Lead foil keel

Bottom of mast

Extend the mast below the hull to support the keel

Does the mast have to be round? Must it be vertical? How many masts can one use? If one mast is good, would two masts be twice as good? How about three? Try and see. Your own hull is your best authority when you want answers to questions like these.

What material is best for your mast? Toothpicks are commonly used as masts on cork hulls; but, are they best? Wood swab sticks and broom straws are longer. When does a mast become too tall? Ask your boat. Copper wire and paper-clip wire have been used for masts. The lower end of the wire can be made into the framework for the keel. Because it is hard to attach keels to these small boats, it is suggested that you extend the mast below the hull to support the keel.

CHANGE THE MAST

Should the mast be at the stern so that the sails can push the boat like a motor? Should it be centered so that the boat will turn more easily? Or should the mast be at the bow of the boat to pull the craft along?

Using a split hull makes it very easy to experiment with the position of the mast. You can use more than one rubber band to hold the mast, the keel, and the rudder more securely.

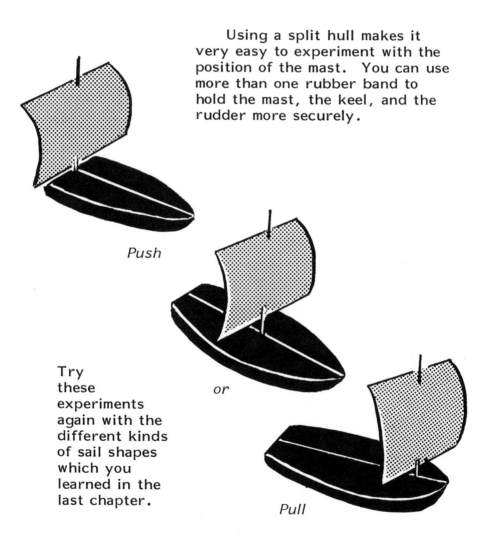

Push

or

Try
these
experiments
again with the
different kinds
of sail shapes
which you
learned in the
last chapter.

Pull

CHANGE THE KEEL

50¢ coin keel

A keel should be heavy. If aluminum foil is the only flat metal available to you, fold it many times or wrap a heavy nail or bolt at the bottom. Single-edge razor blades have been used with care on these small boats. Does the keel have to be curved and beautifully shaped to sail straight and fast, or is that just for looks?

The keel of your sailboat should be heavy enough to right the mast if it is blown over by the wind. This should happen even if the sail is wet and heavy. Plumbers' sheet lead is best for your keel. If you cannot get this, lead foil and bottle-seal foil are your next best keel materials. Copper coins have often been used for keels on these small boats.

The heavy keel must overbalance the mast and the sail

CHANGE THE SAIL

Experimenting with sails is the most fun of all. In these coming days of energy shortage, sailboats will never lack power. Would more than one sail give your boat more power and speed? Why don't you try and see?

Fasten your sails to the masts with glue or thread. Any wood which is used at the bottom of the sail for stiffening, for strengthening, or for weighting is known as a boom. Any wood used across the top of the sail is called a spar.

Change the size. If you made a sail twice as large, would it be more efficient than two sails? Would three small sails be better than your single sail of equal area? Your own boats will give you the best answers to specific questions such as these.

Change the material. Use wax paper, aluminum foil, or Saran Wrap for sail materials to start. Don't overlook the thin sheet foam such as is used for egg cartons and food packs.

Change the shape. Should your sail be tall and thin or short and wide? Should it be triangular, square, or some other shape? What sail is best for your boat?

Lateen sail on an Egyptian felucca

CHANGE THE SAIL

There are many combinations of sails to be seen on sailboats. These are some of the most popular kinds of Marconi-rigged sailboats.

Can you combine square, triangular, and lateen sails? You certainly can. You can try anything when you are experimenting.

Sloop

Yawl

Ketch

Any boat will sail with the wind. Can you make a boat that will "tack" into the wind? A boat that is beating to windward zig-zags back and forth into the wind.

Schooner

CHANGE THE RUDDER

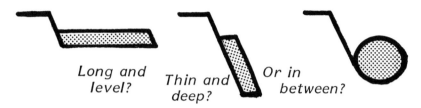

Long and level? Thin and deep? Or in between?

Which rudder shape is best?

A good sailboat should sail straight with the wind all by itself. A rudder can correct some imbalance, but its real use is for turning the boat.

The simplest rudder for your small boat can be made quickly by bending a paper-clip wire or a copper wire into a combination rudder, post, and tiller. Be sure to try some of your own designs to find what works best for your boat. What shape is best for the rudder? What size is best? What material is best? What weight is best? Should the rudder be long and horizontal, thin and deep, or round and in between? You will enjoy finding out for yourself what is best for your particular sailboat.

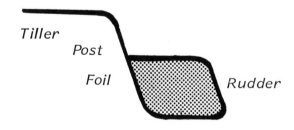

Tiller

Post

Foil

Rudder

The simplest paper-clip wire rudder

MAKE YOUR OWN POOL

It is easy to improvise a shallow pool on the ground, on the floor, or even on a long table. Buy one of the backpackers' tube tents at a sporting goods store. Slip this over a rectangular frame which has been made with long 2" x 4" boards. Smooth these boards well with sandpaper so that slivers cannot cause leaks. Fold the open ends of the plastic tube under the end boards and tape them into place. Fill your pool with a garden hose and put an electric fan at one end. This pool is ideal for races as well as for experimenting.

Fan

Backpackers'
tube tent

Siphon
the water
out of the
pool with the
same garden
hose when you
are finished.

Tape
the
ends

A VARIATION

A Kontiki stamp boat

Young children have sailed stamp boats as long as cork has been available. Teenagers, too, enjoy racing these little boats. Anyone can make one, but can you make one that will sail straight and fast? The urethane sheet foam from egg cartons is as good as cork for making an easy-to-cut, raft-like hull. Why don't you organize a Kontiki Regatta?

Bubble boats are fun too. They are made from the plastic foam half-bubbles which are used for shock-proofing some packages for mailing. A leaf or a stamp can be the sail. Broom straws, toothpicks, and pins make good masts. A shingle nail can be a good rudder.

Half-bubble boat

ANOTHER VARIATION

Diagonally-cut paper milk cartons are ready-made hulls. Stick the pencil mast into a triangular block of plastic foam and add more than the usual amount of ballast (weight) in the hull and keel. This hull is good for experimenting with outriggers and catamarans. It is large enough to sail well in a swimming pool or on a lake. Try all of your experiments over again on each of these new variations. Try predicting what will happen with each change from your experience with the cork boats.

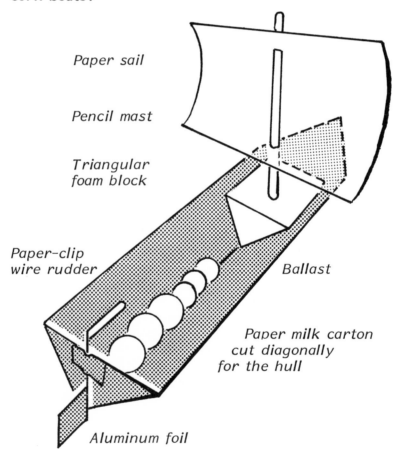

Paper sail

Pencil mast

Triangular
foam block

Paper-clip
wire rudder

Ballast

Paper milk carton
cut diagonally
for the hull

Aluminum foil

STILL ANOTHER VARIATION

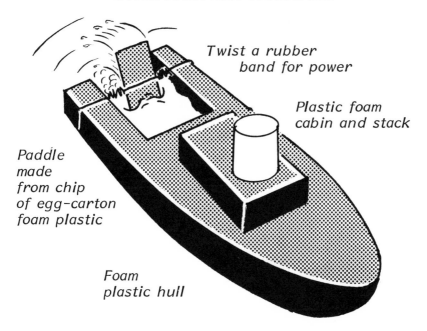

Twist a rubber band for power

Plastic foam cabin and stack

Paddle made from chip of egg-carton foam plastic

Foam plastic hull

The paddle-wheeler

Cut the hull from foam plastic so that you can do much experimenting with the shape. Experiment, too, with the shape of the power paddle. What other experiments of your own can you design?

Young children often enjoy racing rubber-band powered, forked tree banches. If these are dry, they will float higher in the water and the propellers will work better.

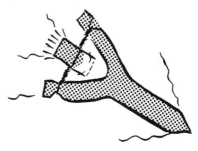

Forked tree branch

AND, AN AMBITIOUS VARIATION

If you have a swimming pool or a larger body of water available, it would be a shame not to experiment with carbon dioxide-propelled rocket boats. You can purchase the carbon dioxide cartridges at hobby shops. Place the cabin to the rear of the boat so that you will have a double thickness in which to cut or drill the hole for the power rocket.

If the water is too rough, stretch a nylon fishing line over the course. The boat will follow this if guide loops are fastened to the bottom.

The ultimate, of course, is to build a radio controlled motorboat. After experimenting with cork boats, you will appreciate more any boat that you build or buy.

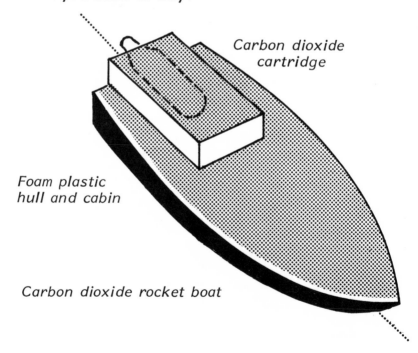

Carbon dioxide cartridge

Foam plastic hull and cabin

Carbon dioxide rocket boat

EXPERIMENTAL FUN WITH THE

INVERTED TUMBLER

HOW MUCH WEIGHT WILL YOURS SUPPORT?

THE PROBLEM

Most people feel that they have achieved something if they can turn over a glass of water and the loose cover stays on. One of the author's science students was amazed, as was his class and his teacher, when his peanut butter jar and loose plastic cover supported seven gallon cans filled with water (thirty-five kilograms). Another student was able to hang twenty-four kilograms from a flat-edged dog food bowl with a "plumbers' friend" for a cover. How much weight can you hold on the loose cover of an inverted container?

MATERIALS NEEDED

Gather as many different kinds of containers as you can besides plastic and glass drinking tumblers. Try many different kinds of materials. Be sure to have different shapes. Be especially observant of the shape of the openings. Be sure that you have flat as well as round containers.

Plasticized milk-carton paper is an excellent cover material as are plastic, snap-on jar tops. Try tin can lids, screens, and even corrugated lids. What you see will amaze you. The hooks on your covers can be bent pins, bent nails, or eye-screw bolts.

Weights can be any objects which you can hang or tape to the covers. Coins are good weights when hung on tape or placed in the container. Fishing sinkers are easy to hang. Plastic bags and plastic medicine bottles, filled with sand or water, are good for heavy weights. Mark or calibrate all of your weights in the metric system.

CONSIDER THIS

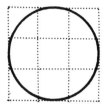

Seven square inches

What keeps the water from spilling out of the inverted tumbler? Most people are quick to say that air pressure does it. They explain that air presses about 14.7 pounds on each square inch of the earth's surface, and that does it. Does it? Consider the following and then go ahead with your experimenting.

The average drinking tumbler has about seven square inches of opening because it is usually three inches in diameter. The air pressure on this area of the cover alone should be seven times 14.7 pounds. This means that 102.9 pounds of air pressure is holding on the cover. Of course, it should stay and not spill.

But wait a minute. Can these same people who explain this so easily put one hundred pounds of pennies on the cover? Can they put fifty pounds? Ten pounds? Or just one little pound? Most of these easy explainers fail when only a penny is added to their cover. A few of these people haven't been able to keep the cover on with no extra weight. Here is a real problem for you.

"Air pressure does it!"

71

EXPERIMENT FIRST WITH

You can learn much with a secured cover in your introductory experimenting with this problem. Cover two tumblers with nylon stocking material. Secure them with rubber bands. Pour water from one tumbler into the other several times.

Then cover the tumbler with your hand and turn it over. Practice over a sink or some other catch basin. When you can do this well, practice turning over the tumbler of water in the air without covering it with your hand. Practice until you can entertain your friends without spilling a drop.

A NYLON-COVERED TUMBLER

Change the attitude
of the tumbler. At what
angle does the water
start to run? Stop?

Change the size of
the holes by poking with
a very sharp pencil.

*Change the attitude
or angle*

Then push a dome
with your finger under
the inverted tumbler as
you hold it with no rub-
ber band. Push the tum-
bler down as you hold
the nylon cover with
your other hand. The
dome will flatten as it
lets in bubbles of air.
This is how the old
Gypsy "boils" water
with her silk 'kerchief
in the carnivals.

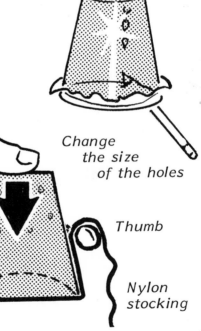

*Change
the size
of the holes*

*Press down
between fingers*

Thumb

Forefinger

*Nylon
stocking*

You can "boil" water like the old Gypsy does

WHAT CAN YOU CHANGE?

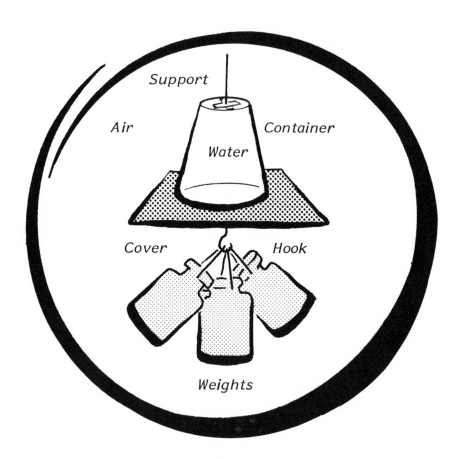

Support

Air

Container

Water

Cover

Hook

Weights

Experimenting with a loose cover is just a little harder than changing things with a secured cover. An easy way to plan experiments is to sketch the setup for working with your problem. Then draw a circle around all the objects and materials in the experimental system. Concentrate on the changes which you can make on these objects and materials.

WHAT CAN YOU CHANGE?

Here are a few suggested changes to get you started. Each suggestion will lead you to many more that you can design into experiments by yourself.

Change the container
    Shape, size, weight, and material

Change the cover
    Shape, size, weight, and material
    Try a corrugated cover
    Put holes in the cover
    Make a nylon stocking cover

Change the water
    Try liquids of other specific gravities
    Use liquids of different viscosities
    Fill with different amounts of liquid

Change the weight
    Shape, size, amount, and material
    Try hanging it differently
    Substitute a spring scale

Change the support
    Hold the container in hand
    Tape the container to a string
    Support the container from the cover

Change the air
    Try this in a vacuum chamber
    Make a vacuum inside the container
    Take the experiment into the
        swimming pool

CHANGE THE CONTAINER

Multiply the number of containers. Could you make one tumbler of water hold to another?

Change the size of the container. What is the smallest container that will work? The largest?

Multiply the number

Change the shape of the container. Be sure that you experiment with changes in the shape of the opening of the container. Does the amount of water directly over the opening make a difference? You can make a long, narrow container by plugging the end of a plastic tube.

Change the size

Change the material of the containers. Tumblers are made of glass, plastic, ceramic, clay, metal, and other materials.

Change the shape   *Change the size of opening*

CHANGE THE CONTAINER

Remember to keep control by changing only one variable at a time. This is easier to say than to do.

Multiply the number

When you change the size, keep material, shape, cover, weight, etc. under control. Change only that one condition. When you change material, keep size, shape, weight, etc. under control.

Sometimes you will not be able to afford this much control. Always do the best that you can with what you have.

Use caution when using glass containers or when changing to an extreme size of the container.

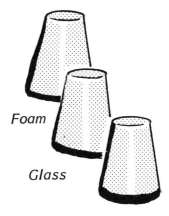

Foam

Glass

Aluminum

"Wow! I did it."

Soft drinks

Peanut butter

Tuna

Change the shape

CHANGE THE SIZE

Thimble

Gallon can

What is the smallest container that will hold up a cover? What is largest?

Five-gallon can with plastic disposal can lid

Size is a condition. A condition is a state of being. It is a mode of existence. It is one of the more-or-less things about an experiment. It can be changed during the event, and it can be changed between a series of related events. A condition is a variable. Most of the changes which you will make in your experimenting will be changes of conditions.

CHANGE THE SIZE OF THE OPENING

Narrow-mouth gallon jar

Challenge:

Write down a prediction. Which of these two gallon containers will hold the most weight on its cover? Perhaps you would even like to say ahead how much weight each will hold.

You will have greater control of the variables if you can get a short, wide tube with a large opening and a tall, thin tube with a narrow opening. The jug and the jar are not quite the same shape which means that you have lost some control. Look carefully at them before you make your prediction.

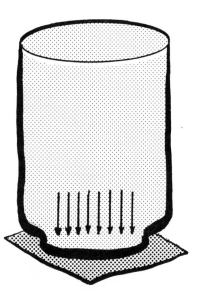

Wide-mouth gallon jar

CHANGE THE ATTITUDE OF THE CONTAINER

Lift the inverted tumbler off the table and turn it slowly in a complete circle. At what part of the turn does it seem to be in the most vulnerable position for a spill? There are 360 degrees in a circle. Identify the tumbler's position in degrees.

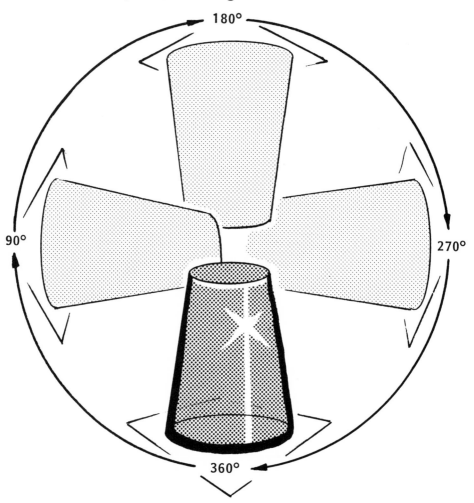

CHANGE THE RIGIDNESS OF THE CONTAINER

"*You're going to get dunked.*"

Plastic bag of water

"**I'll be sitting here long after you slide off that bag.**"

Tuna can

Milk-carton paper cover

    Substitute a plastic bag for the drinking tumbler. You will have to stiffen the neck of the bag to attach a cover. This can be done by removing both ends of a pet food can or tuna can.

CHANGE THE COVER

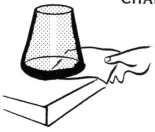

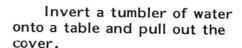
Pull

Invert a tumbler of water onto a table and pull out the cover.

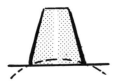

Press

Try a springy cover. Press the cover up in the center and then release it. Remember the dome under the nylon cover? Can you make this happen with a springy cover? Measure the weight-holding ability of the cover before and after pressing the dome.

Balloon

Would a balloon serve as a cover? Try and see. What are its advantages and disadvantages? Would it be best to put hot water or to put cold water in the container?

Corrugated cover

Substitute other materials for the cover. Try wax paper, blotting paper, tissue paper, corrugated paper, plastic screen, metal screen, gauze, and cotton cloth.

CHANGE THE COVER

Change the shape of the cover. Would a square, flat-round, or ball shape be best?

Would an egg or a ball serve as a cover? How much water would work best?

Try a light bulb. It has a handy end on which to tape weights.

Who said that the cover had to be flat? What shape would be ideal?

Poke holes in the cover with a sharp pencil point.

CHANGE TO AN ELASTIC COVER

Attach a "plumbers' friend" to a wide-mouth, flat-rimmed container. Could you fill both the container and the elastic cover with water and place the two amounts together without spilling a drop?

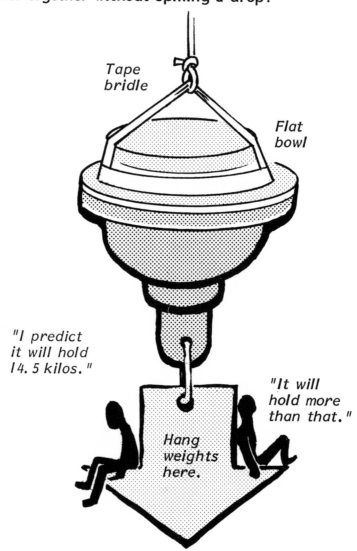

Tape
bridle

Flat
bowl

"I predict
it will hold
14. 5 kilos. "

"It will
hold more
than that. "

Hang
weights
here.

Tape bridle

This is a cross-section drawing.

"Oh, yes it will."

"It won't work."

Hang weights here.

Ask your mother if, in doing her dishes, she has ever put one hot, wet tumbler over another inverted tumbler standing in her drying rack. What happened? Try to find out why. Use very hot water so that the tumblers themselves can get very hot and wet.

CHANGE TO AN ELASTIC CONTAINER
WITH AN ELASTIC COVER

Invite a friend to share
your experimental fun.

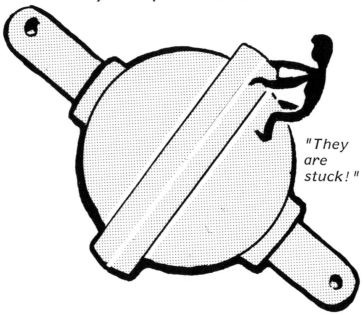

*"They
are
stuck!"*

Compare rigid and elastic materials. Your own
apparatus will tell you what works best better than
books, but books can add data after you get the
feel of the experimental conditions. It is best to
ride a horse and get a sore seat before you read an
equestrian book. Get your feet wet and cough
some water up your nose before you read that book
about how to be an Olympic swimmer.

CHANGE TO A MORE ELASTIC CONTAINER

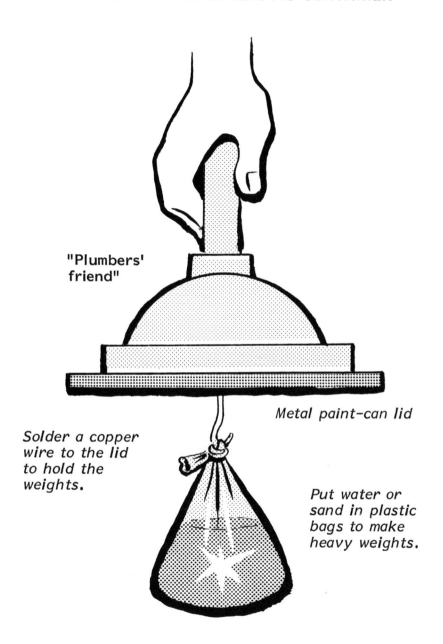

"Plumbers' friend"

Metal paint-can lid

Solder a copper wire to the lid to hold the weights.

Put water or sand in plastic bags to make heavy weights.

CHANGE THE COVER

Try a car-top suction cup on an appropriate jar lid. Also try it on the jar.

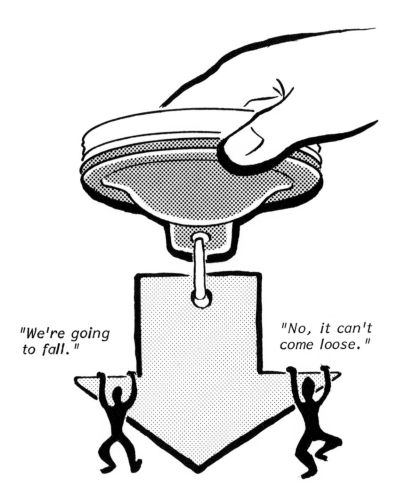

What is your prediction?

CHANGE THE COVER

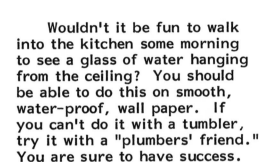

Wouldn't it be fun to walk into the kitchen some morning to see a glass of water hanging from the ceiling? You should be able to do this on smooth, water-proof, wall paper. If you can't do it with a tumbler, try it with a "plumbers' friend." You are sure to have success.

Could you hang a plastic tumbler of water from the cover as shown below? Cover it, twist it, and watch it spin around and around. Next, swing it like a pendulum.

Bent pin or bent nail taped to milk-carton paper cover

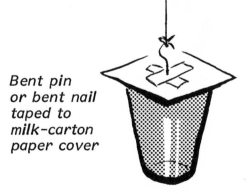

CHANGE THE ATTITUDE OF THE CONTAINER

Swing it gently like a pendulum in the upside down position. Then add a centrifugal force to gravity by swinging it horizontally, around and around.

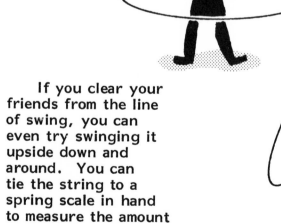

If you clear your friends from the line of swing, you can even try swinging it upside down and around. You can tie the string to a spring scale in hand to measure the amount of force you are adding.

ELIMINATE THE CONTAINER

Fill the container with water, invert it, and
place it in the refrigerator to freeze. Warm the
container with your hands to take it off the
frozen water. How long will the cover cling
to the frozen water?

Be sure to use a cover which is not affected
by soaking. Milk-carton paper has a water-proof,
plasticized surface. Plastic, snap-on jar covers,
also, should be tried.

Are you still predicting the outcome of each
new experiment? This is an easy way to test
your understanding of this phenomenon.

*Will the frozen
water hold the
cover?*

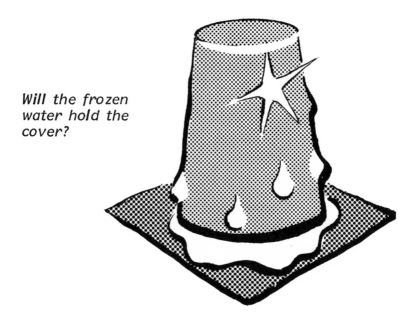

CHANGE THE WEIGHT

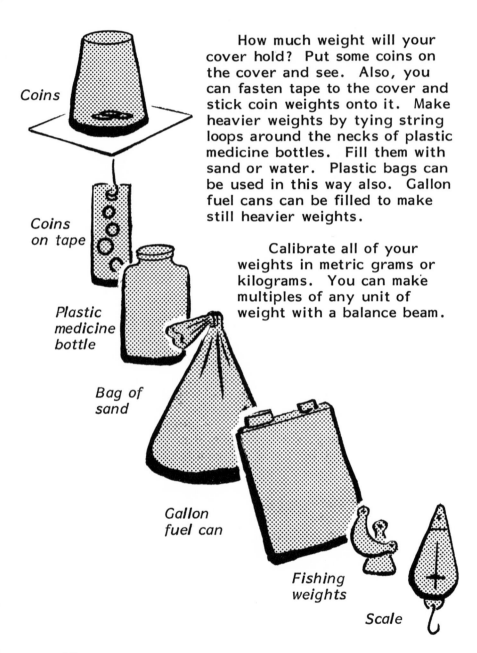

How much weight will your cover hold? Put some coins on the cover and see. Also, you can fasten tape to the cover and stick coin weights onto it. Make heavier weights by tying string loops around the necks of plastic medicine bottles. Fill them with sand or water. Plastic bags can be used in this way also. Gallon fuel cans can be filled to make still heavier weights.

Calibrate all of your weights in metric grams or kilograms. You can make multiples of any unit of weight with a balance beam.

Coins

Coins on tape

Plastic medicine bottle

Bag of sand

Gallon fuel can

Fishing weights

Scale

CHANGE THE LIQUID WITH ADDITIVES

Add soap to a half tumbler of water and shake. Also try detergent as an additive.

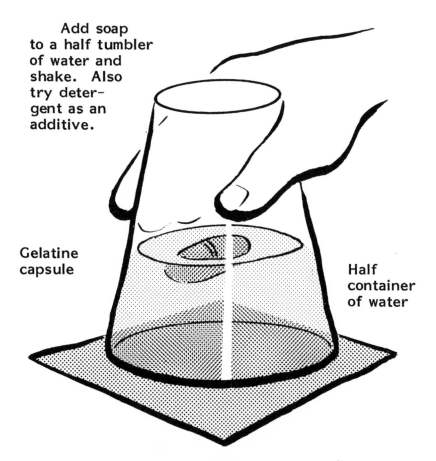

Gelatine capsule

Half container of water

Purchase a large gelatine capsule at the drug store. Fill it with baking soda and drop it into a half-glass of lemon juice. Cover the tumbler quickly and invert it. Also try twisting a small piece of paper around your little finger to make a container to fill with baking soda. Drop this package into a half-glass of vinegar. Cover the tumbler quickly and invert it.

CHANGE THE LIQUID

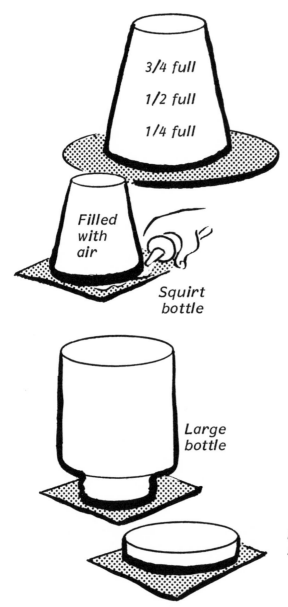

3/4 full

1/2 full

1/4 full

Filled with air

Squirt bottle

Large bottle

Lid of the same bottle

Change the amount of water in the container. Even try a container empty of water but sprinkle a little water on the cover around the rim of the container.

Change the the amount of the liquid over the opening of the container. Using the cover for a container gives you good control over the size of the opening. Keep all factors under control except the one which you are manipulating.

Are you still going to extremes first? You find out faster what you want to know that way.

CHANGE THE LIQUID

Compare the weight-holding ability of boiling water and ice water. If you put both these inverted tumblers of water in the refrigerator freezing compartment, which will freeze first? You won't believe what you see. Try it several times. How do you explain this? Be sure that you keep control of all the variables. After each has been frozen, which will hold the most weight hooked to the cover?

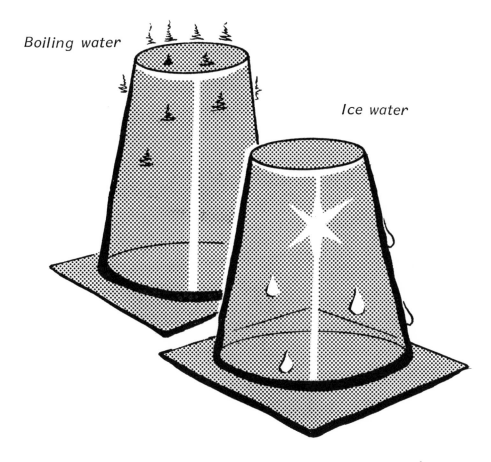

Boiling water

Ice water

CHANGE THE LIQUID

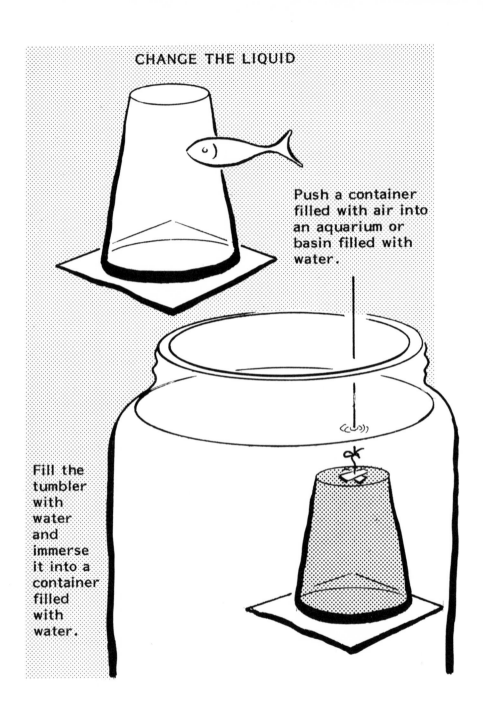

Push a container filled with air into an aquarium or basin filled with water.

Fill the tumbler with water and immerse it into a container filled with water.

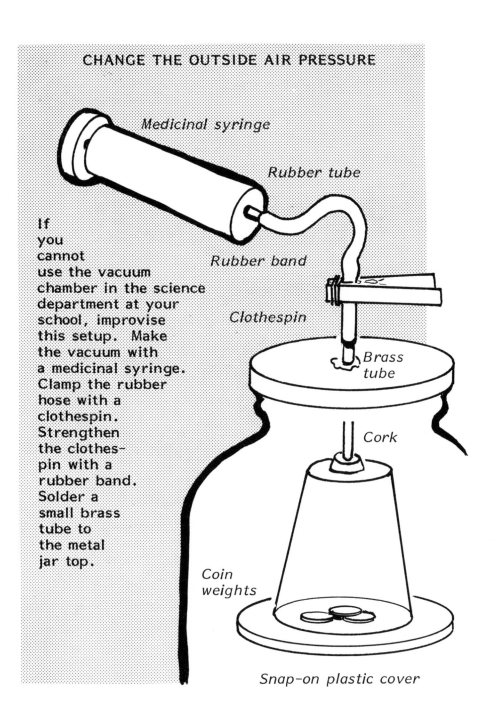

CHANGE THE OUTSIDE AIR PRESSURE

Medicinal syringe

Rubber tube

Rubber band

Clothespin

Brass tube

Cork

If you cannot use the vacuum chamber in the science department at your school, improvise this setup. Make the vacuum with a medicinal syringe. Clamp the rubber hose with a clothespin. Strengthen the clothespin with a rubber band. Solder a small brass tube to the metal jar top.

Coin weights

Snap-on plastic cover

CHANGE THE INSIDE AIR PRESSURE

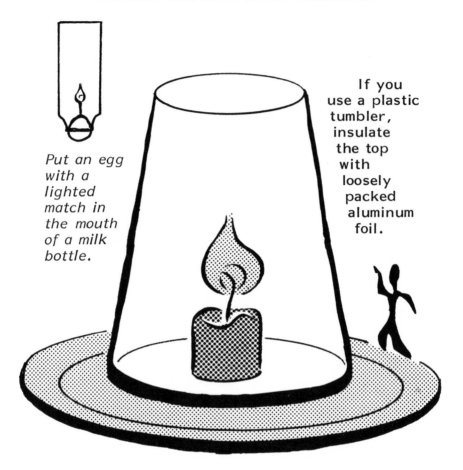

Put an egg with a lighted match in the mouth of a milk bottle.

If you use a plastic tumbler, insulate the top with loosely packed aluminum foil.

Fasten a burning, stub candle to a round plastic cover with a few drops of the hot wax. Put the round cover in a plate of water. Then place an inverted drinking tumbler of air over the burning candle. The candle will burn the oxygen in the air. A vacuum will form in its place.

CHANGE THE INSIDE AIR PRESSURE

Steam

Plate of ice water

Fill a plastic tumbler with steam. Press it onto a plastic cover floating on a plate of ice water.

Then put a plastic cover on a one-third full glass of hot water, invert it, and put it in the refrigerator to cool.

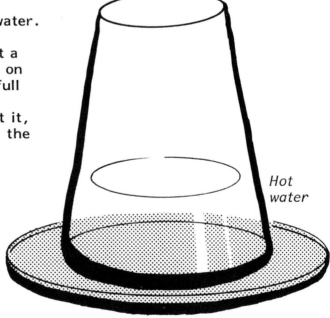

Hot water

CHANGE THE DIRECTION OF THE FORCE

Shake the container up and down by hand, slowly at first and then harder and harder. Be sure that you have a catch basin under the inverted tumbler. Then shake it so hard that the cover comes off, spilling the water.

Rubber band

Put a tape bridle on a dry container and support the water-filled, inverted tumbler on a strong rubber band with an "S" hook. Give it a push so that it bounces up and down. Be sure to add more and more weight to the cover.

EXPERIMENTAL FUN WITH THE

CARTESIAN DIVER

HOW FAST CAN
YOU RECOVER
THE TREASURE?

THE PROBLEM

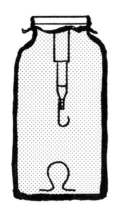

Cartesian diver

Use a Cartesian diver to see how fast you can recover the treasure (a wire ring) from the bottom of a bottle or jar. Glue, tie, or tape a thin wire hook to the end of a medicine dropper. Squeeze air from the dropper until it barely floats. Cover the bottle with rubber from a balloon. Press on this cover, and your submarine will dive to lift the treasure. Practice until you can control it and understand how it works

MATERIALS NEEDED

Containers	Bottles and jars of many shapes, sizes, and materials
Covers	Rubber balloons, thin plastic sheets, and rubber bands
Divers	Medicine droppers, syringes, test tubes, vials, and small bottles
Hooks	Thin copper wire, pins, paper clips, thread, and nylon fishing line
Liquids and solutions	Water, oil, syrup, soap, detergent, gelatin, sugar, and salt

CHANGE THE COVER

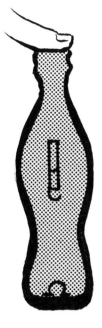

Press on a narrow-neck bottle with a finger.

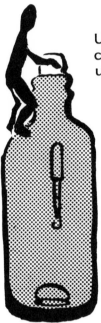

Use a cork for a cover and push it up and down.

Put your palm over larger openings to press.

Try toy balloon rubber for a cover. Use a rubber band to keep it in place. Also try the sheet plastic from a polyethylene clothing bag.

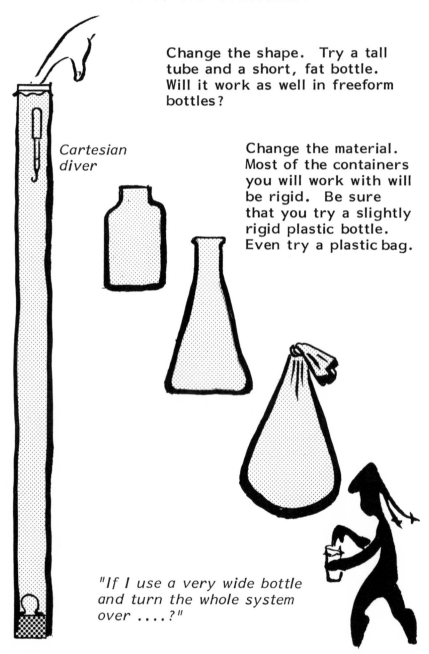

Cartesian diver

Change the shape. Try a tall tube and a short, fat bottle. Will it work as well in freeform bottles?

Change the material. Most of the containers you will work with will be rigid. Be sure that you try a slightly rigid plastic bottle. Even try a plastic bag.

"If I use a very wide bottle and turn the whole system over ....?"

CHANGE THE CONTAINER

What will happen if you turned the container over and pressed up instead of down? Would this force the diver to the surface? Try and see. Be sure that the container is wider than the diver is long so that the diver is free to move in the water.

What new experiments can you design now to learn more about the Cartesian diver? Perhaps you are wondering why this phenomenon is called the Cartesian diver. It is name after Rene Descartes, a famous French philosopher, mathematician, and scientist. He discovered this submarine principle in the early 1600s. It became the submarine, and it has been reproduced in many forms as a diving toy to entertain ever since that time.

"I don't believe it!" *"What do you see?"* *"Hey, let me see."*

Invert the container and press upward

CHANGE THE DIVER

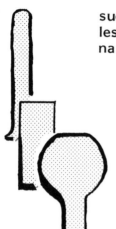

The diver must barely float to be successful. Experiment with more and less air in it. This is hard to do in a narrow-neck container.

Often it is best to adjust the diver in a wide-mouth jar, pan, or basin. Transfer it to the narrow-mouth container without losing a drop of water when you are ready to start the dives.

What would happen if you plugged the opening of the medicine dropper or other kind of diver?

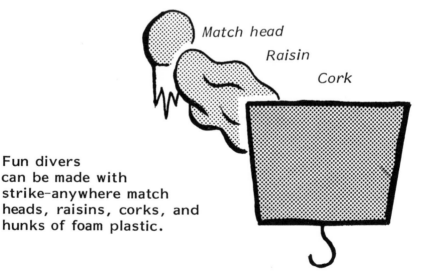

Match head

Raisin

Cork

Fun divers can be made with strike-anywhere match heads, raisins, corks, and hunks of foam plastic.

CHANGE THE HOOK ON THE DIVER

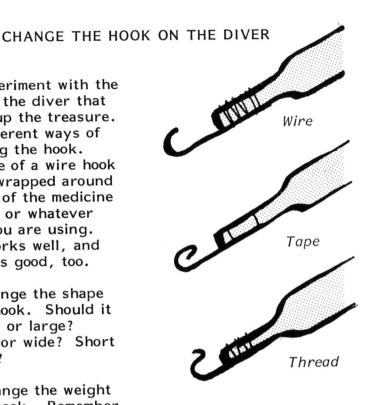

Wire

Tape

Thread

Experiment with the hook on the diver that brings up the treasure. Try different ways of fastening the hook. The wire of a wire hook can be wrapped around the end of the medicine dropper or whatever diver you are using. Tape works well, and thread is good, too.

Change the shape of the hook. Should it be small or large? Narrow or wide? Short or long?

Change the weight of the hook. Remember that the diver must barely float to be most successful. You may have to add much weight to cork or to foam plastic to get this condition.

Watch the water level in the diver very carefully because it will be trying to tell you something.

Change the shape

It was suggested at the beginning of this chapter that you start by using a medicine dropper for a diver because it was so easy to fill and to float at the right level. As you begin using other divers, the problem of filling them becomes difficult. A wide container makes it easier, but many times you will want to experiment with a narrow-mouth container. A flexible, bent soda straw can help.

CHANGE THE AIR

Change the amount
of air above the water
in the container. Does
the container have
to be full?

1/4 full

1/2 full

3/4 f

Make the diver
airtight with a
cork, stopper,
or screw lid.

Fill the diver
with a lighter-
than-water
liquid to re-
place the air.

Use other materials than air
in the diver and even in the
large container.

Rubbing alcohol

CHANGE THE AIR PRESSURE

Blow more air pressure into the container with your mouth. Blow on the bottle with and without the rubber or the plastic cover.

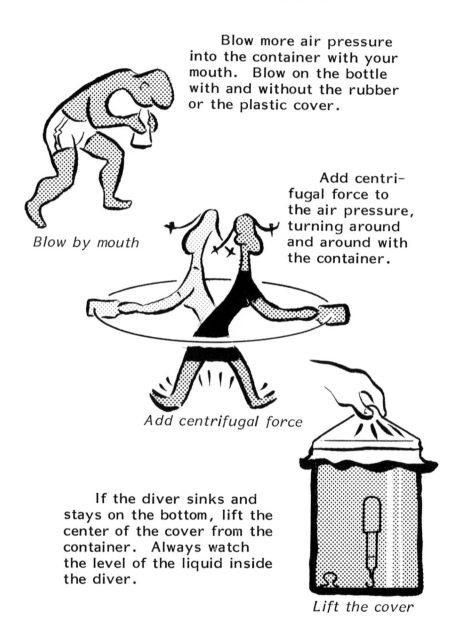

Blow by mouth

Add centrifugal force to the air pressure, turning around and around with the container.

Add centrifugal force

If the diver sinks and stays on the bottom, lift the center of the cover from the container. Always watch the level of the liquid inside the diver.

Lift the cover

CHANGE THE AIR PRESSURE

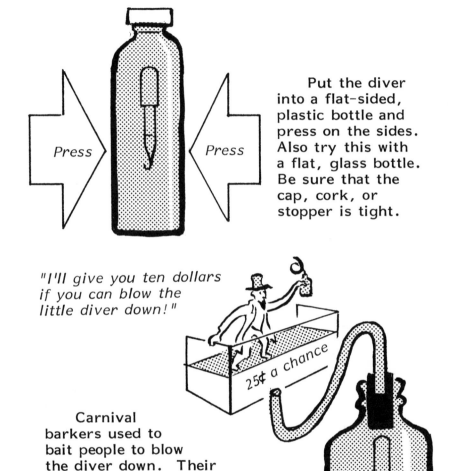

Put the diver into a flat-sided, plastic bottle and press on the sides. Also try this with a flat, glass bottle. Be sure that the cap, cork, or stopper is tight.

"I'll give you ten dollars if you can blow the little diver down!"

25¢ a chance

Carnival barkers used to bait people to blow the diver down. Their tubes were plugged at the end. Few people knew that the barker was really sending the diver down by pressing the sides of the flat bottle while he was misleading them by blowing on the tube.

CHANGE THE LIQUID

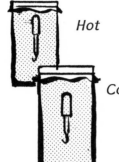

Hot

Cold

Change to other
liquids and solutions
such as:

Soda pop
Vinegar
Mineral oil
Syrup
Molasses
Honey

Use additives in
the water:

Salt   *Very*
Sugar   *long*
Soap   *plastic*
Gelatin   *tube*
Detergent

If we changed the depth
of the liquid with a long
plastic tube ....?

*"Hey, press harder
up there."*

EXPERIMENTAL FUN WITH THE

HANG GLIDER

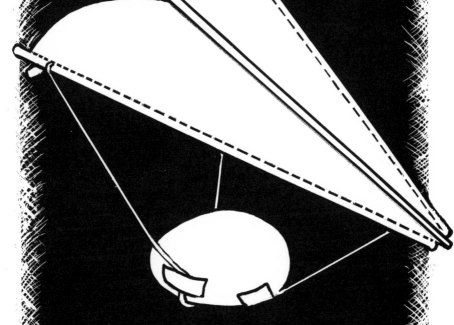

CAN YOU LAND AN EGG SAFELY?

THE PROBLEM

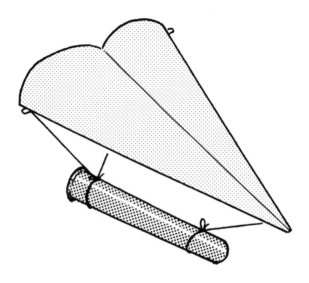

The Rogallo hang glider

Can you sail a Rogallo hang glider farthest while bringing down an egg or a glass test tube unbroken? Rogallo hang gliders are the easiest to make and the simplest to fly. You can learn much about the large ones by sailing miniature gliders. Four sticks or straws, a thin sheet, and a sling underneath for the payload are all that are needed. You must discover the best way to put these materials together to do the job. The simplest and most common arrangement is shown above. It is a good way to start; but remember, you can change sheets, struts, sling, and even the air to improve it.

SOME GOOD ADVICE

As with the sailboat and the wind wagon, young experimenters often get so interested and so excited about hang gliders that they want to make a large one right away and ride it themselves. Please don't make a large glider right away just to jump off the garage roof – on your way to the hospital. It won't work.

When you think you are ready to take to the air yourself, enroll in a hang glider school to have your glider checked out before leaving the ground. Best of all, learn the fundamentals of gliding, aerophysics, tools of the trade, and safety techniques. Keep yourself in good health so that you can enjoy your hang gliders. Get as much experience as you can with model hang gliders before you spend your time and money on a large model and the schooling to fly it.

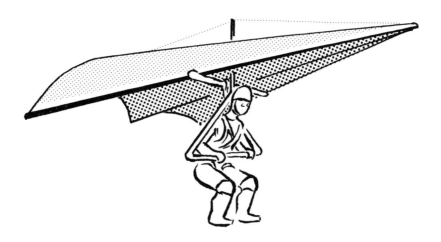

You, too, can fly like a bird after you learn how birds fly.

PARTS OF A ROGALLO HANG GLIDER

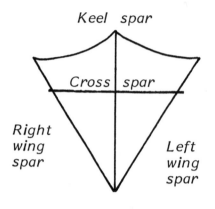

Keel spar

Cross spar

Right wing spar

Left wing spar

Top view

Make a soda-straw size glider first. Broom bristles, swab sticks, and balsa wood sticks can also be used. Fasten the spars with thread or tape at first. Then use glue after you have decided which conditions go together best.

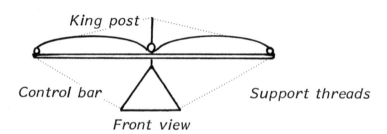

King post

Control bar

Support threads

Front view

The king post and supporting threads are not necessary on small gliders with light-weight payloads.

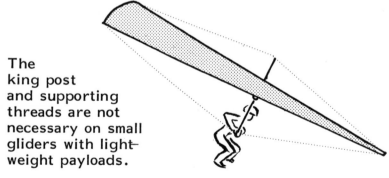

Side view

CHANGE THE SPARS (RODS OR TUBES)

Changing the relationship of your spars will alter the shape of your glider fastest, especially the changes in the placement of the cross spar.

In which position does the cross spar work best? Forward, backward, or middle?

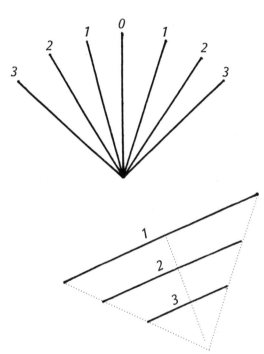

Do all spars have to be the same length? Could the keel be longer or shorter? Slit the ends of the soda straws and splice them together end to end to make longer straws.

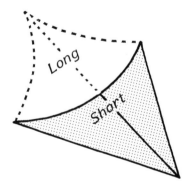

CHANGE THE SAIL (SHEET)

The simplest way to fasten the sail

Fasten the sail with rubber cement, contact glue, pins, or thread. Fasten it so that you can change it fast. If the sail is fastened to itself rather than to the wing spars, it can be slipped on and off more easily. The payload can be fastened with electronic mini clamps while you are experimenting with the sail.

Try different materials. Saran Wrap is a good sail material for small model gliders. The thin polyethylene clothing bag material is also good. Tissue paper and many other lightweight sheets will work.

Don't overlook the possibilities of a stiff-wing glider made from egg-carton tops. Glue two of the foam sheets down the middle. Experiment with the amount of tilt that is best for your wings. This tilt angle is known as the dihedral angle.

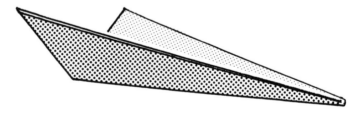

Stiff-wing glider made from egg-carton foam

CHANGE THE GUIDANCE

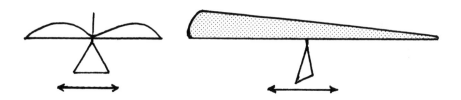

Turns can be made by shifting the payload from side to side.

Dives and stalls can be controlled by shifting the payload weight forward or backward.

Fins for guidance can be added above or below the ends of the spars.

Experiment by lowering the keel spar under the cross spar to use it for guidance and to increase the dihedral angle.

Read about the dihedral angle in model building books

CHANGE THE CONTROL BAR

The control bar is a triangular trapeze arrangement on which the rider hangs and shifts his weight
on large hang gliders. You can bend a control bar
from a large paper clip or from copper wire to send
a small doll aloft. Coat hanger wire is used for larger
models. Remember, the king post which extends above
the control bar is not necessary for light payloads.

A payload often can be landed better with an
almost weightless thread bridle to replace the control
bar. Hang the payload on three threads from the
corners of the glider. A well-balanced glider with a
good dihedral angle can be flown with a one-thread
bridle hung from the two ends of the keel spar. A
hang glider which is rigged in this way should fly
as a kite if it is well balanced. This is why hang
gliders are often called kites.

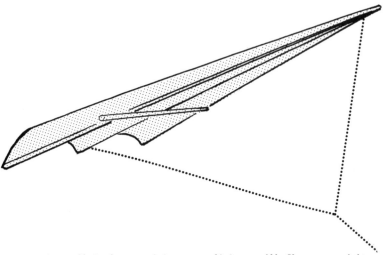

A well-balanced hang glider will fly as a kite

CHANGE THE PAYLOAD

Eggs, glass test tubes, and small dolls are good payloads for testing these miniature gliders. Should the egg be full or empty? You can drain an egg by putting pin holes in both ends and blowing.

Change the size, weight, and shape of the payload. Be sure to experiment with its position. Move it forward and backward. Also give some thought to how far it should be hung below the glider.

You will really be experimenting with the center of mass (often called the center of gravity) when you do this. Read about the center of mass in the encyclopedia and science books at your library.

Because modeling clay can be changed so easily and so fast, it is often used in weight experiments. It is easy to stick onto the bridle strings of the glider. Other payloads can be fastened to the bridle with the smallest electronic clamps.

How about adding shock absorbers to your payload? Expect them to add to the weight. How about making changes in the landing surface?

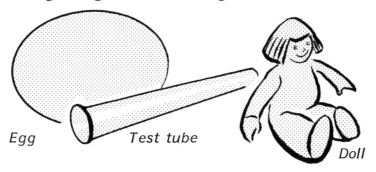

Egg   Test tube   Doll

Good payloads for testing hang glider models

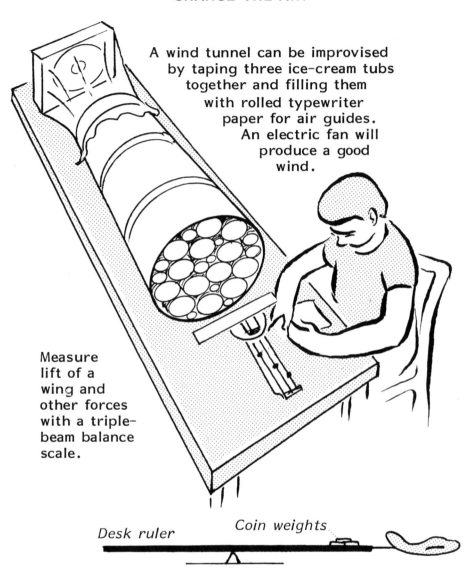

A wind tunnel can be improvised by taping three ice-cream tubs together and filling them with rolled typewriter paper for air guides. An electric fan will produce a good wind.

Measure lift of a wing and other forces with a triple-beam balance scale.

Desk ruler   Coin weights

Improvise a desk ruler into a scale

CHANGE THE LAUNCH METHOD

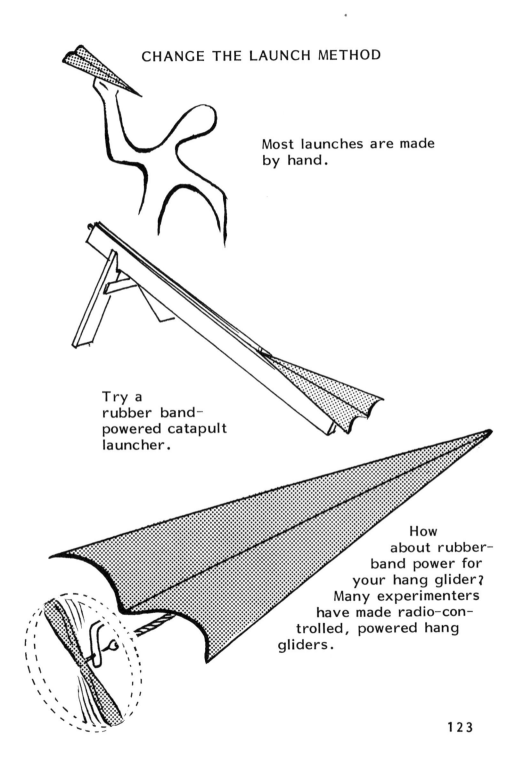

Most launches are made by hand.

Try a rubber band-powered catapult launcher.

How about rubber-band power for your hang glider? Many experimenters have made radio-controlled, powered hang gliders.

CHANGE TO EGG-CARTON FOAM GLIDERS

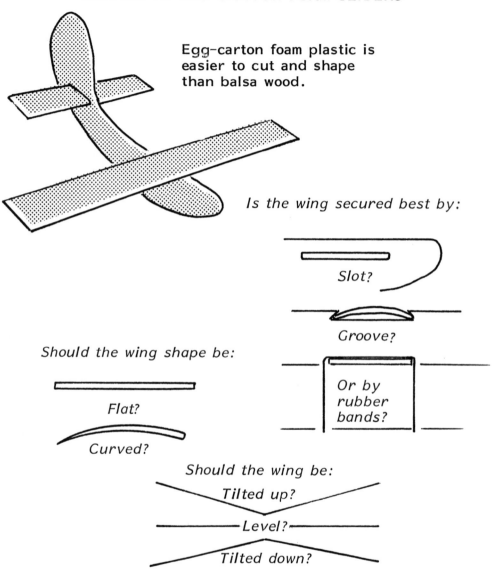

Egg-carton foam plastic is easier to cut and shape than balsa wood.

Is the wing secured best by:

Slot?

Groove?

Should the wing shape be:

Flat?

Curved?

Or by rubber bands?

Should the wing be:

Tilted up?

Level?

Tilted down?

Do your best theories work on the tail also?

EGG-CARTON FOAM GLIDERS

Should the fuselage be:

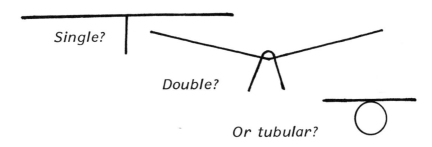

Single?

Double?

Or tubular?

Should the wings be:

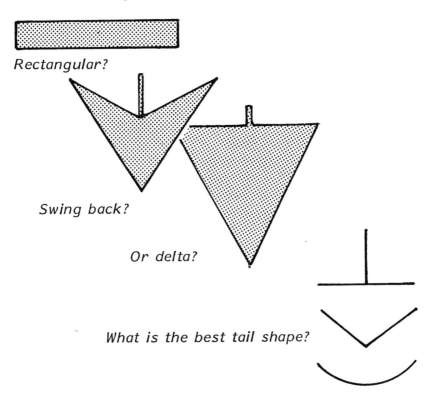

Rectangular?

Swing back?

Or delta?

What is the best tail shape?

CAN YOU MAKE A PAPER GLIDER THAT WILL SAIL FARTHEST?

Make no folds in this area.

Anyone can make a good paper glider since <u>The Great International Paper Airplane Book</u> has been written by Howard Gossage and Jerry Mander. Try the above problem on your friends who are glider buffs. You will find quickly who really understands what makes a glider fly.

One learns most thoroughly by doing; but after you have had some experience, such as experimenting with the problems in this book, is a good time to go to the library to get some books about the things which you have been trying. Ask your librarian to help you find the books listed on the next page and other related material.

BIBLIOGRAPHY

Adkins, Jan — *The Craft of Sail*

Barnaby, Ralph S. — *How to Make and Fly Paper Airplanes*

Gibbs, Tony — *Practical Sailing*

Gossage, Howard and Mander, Jerry — *The Great International Paper Airplane Book*

Henderson, Richard — *Hand, Reef and Steer*

Jaber, William — *Wheels, Boxes, and Skate Boards*

Joseph, Joan — *Folk Toys Around the World and How to Make Them*

Kaufman, John — *Flying Hand-launched Gliders*

Kettelkamp, Larry — *Spinning Tops*

Renner, Al G. — *How to Build a Better Mousetrap Car*

Renner, Al G. — *How to Make and Use a Microlab*

Ritchie, Carson I. A. — *Making Scientific Toys*

Ross, Frank, Jr. — *Flying Paper Airplane Models*

Simon, Seymour — *Projects with Air*

Stensbol, Ottar — *Model Flying Handbook*

Zarchy, Harry — *Let's Go Boating*

INDEX